ENVIRONMENTAL HEALTH – PHYSICAL, CHEMICAL
AND BIOLOGICAL FACTORS

MARINE AND FRESHWATER HARMFUL ALGAL BLOOMS

ENVIRONMENTAL HEALTH – PHYSICAL, CHEMICAL AND BIOLOGICAL FACTORS

Asbestos: Risks, Environment and Impact
Antonio Soto and Gael Salazar (Editors)
2009. ISBN: 978-1-60692-053-4

Environmental Change and Medicine
Viroj Wiwanitkit
2009. ISBN: 978-1-60876-155-5

Phthalates and Bisphenol - A in Plastics and Possible Human Health Effects
Gail N. Moye (Editor)
2009. ISBN: 978-1-60692-802-8
2009. ISBN: 978-1-60876-721-2 (E-book)

Ecological Approaches to Health: Interactions Between Humans and their Environment
Claire Dumont
2009. ISBN: 978-1-60741-061-4

Environmental Regulation: Evaluation, Compliance and Economic Impact
Diederik Meijer and Fillipus De Jong (Editors)
2009. ISBN: 978-1-60741-645-6

Environmental Health Risks: Lead Poisoning and Arsenic Exposure
Jack D. Gosselin and Ike M. Fancher (Editors)
2009. ISBN: 978-1-60741-781-1

Cadmium in the Environment
Reini G. Parvau (Editor)
2010. ISBN: 978-1-60741-934-1

**Cadmium Toxicity
and the Antioxidant System**
*A.C. Pappas, E. Zoidis, K. Fegeros,
G.Zervas and P.F. Surai*
2010. ISBN: 978-1-61668-172-2
2010. ISBN: 978-1-61668-463-1 (E-book)

**Marine and Freshwater
Harmful Algal Blooms**
Peter E. Williams (Editor)
2010. ISBN: 978-1-60741-838-2

**Removal of Toxic Chromium
from Wastewater**
*Tonni Agustiono Kurniawan
and Mika E.T. Sillanpää*
2010. ISBN: 978-1-60876-340-5

**Legionella Pneumophila:
From Environment to Disease**
*Atac Uzel and E. Esin Hames-Kocabas
(Editors)*
2010. ISBN: 978-1-60876-947-6

**The Diversity of Cypriniforms
Throughout Bangladesh: Present Status
and Conservation Challenges***
Mostafa A. R. Hossain and Abdul Wahab
2010. ISBN: 978-1-61668-765-6
2010. ISBN: 978-1-61728-473-1 (E-book)

ENVIRONMENTAL HEALTH – PHYSICAL, CHEMICAL
AND BIOLOGICAL FACTORS

MARINE AND FRESHWATER HARMFUL ALGAL BLOOMS

PETER E. WILLIAMS
EDITOR

Nova Science Publishers, Inc.
New York

Copyright © 2010 by Nova Science Publishers, Inc.

All rights reserved. No part of this book may be reproduced, stored in a retrieval system or transmitted in any form or by any means: electronic, electrostatic, magnetic, tape, mechanical photocopying, recording or otherwise without the written permission of the Publisher.

For permission to use material from this book please contact us:
Telephone 631-231-7269; Fax 631-231-8175
Web Site: http://www.novapublishers.com

NOTICE TO THE READER

The Publisher has taken reasonable care in the preparation of this book, but makes no expressed or implied warranty of any kind and assumes no responsibility for any errors or omissions. No liability is assumed for incidental or consequential damages in connection with or arising out of information contained in this book. The Publisher shall not be liable for any special, consequential, or exemplary damages resulting, in whole or in part, from the readers' use of, or reliance upon, this material. Any parts of this book based on government reports are so indicated and copyright is claimed for those parts to the extent applicable to compilations of such works.

Independent verification should be sought for any data, advice or recommendations contained in this book. In addition, no responsibility is assumed by the publisher for any injury and/or damage to persons or property arising from any methods, products, instructions, ideas or otherwise contained in this publication.

This publication is designed to provide accurate and authoritative information with regard to the subject matter covered herein. It is sold with the clear understanding that the Publisher is not engaged in rendering legal or any other professional services. If legal or any other expert assistance is required, the services of a competent person should be sought. FROM A DECLARATION OF PARTICIPANTS JOINTLY ADOPTED BY A COMMITTEE OF THE AMERICAN BAR ASSOCIATION AND A COMMITTEE OF PUBLISHERS.

LIBRARY OF CONGRESS CATALOGING-IN-PUBLICATION DATA

Marine and freshwater harmful algal blooms / editor: Peter E. Williams.
 p. cm.
 Includes index.
 ISBN 978-1-60741-838-2 (hardcover)
 1. Algal blooms--United States. 2. Algal blooms--Health aspects. 3. Algal blooms--Environmental aspects. 4. Algal blooms--Research--United States. I. Williams, Peter E.
 QK568.B55M37 2009
 579.8'165--dc22
 2009041026

Published by Nova Science Publishers, Inc. ✣ New York

QK
568
.B55
M37
2010

CONTENTS

Preface		ix
List of Acronyms		xvii
Chapter 1	Legislative Background and Report Process *Joint Subcommittee on Ocean Science and Technology*	1
Chapter 2	The HAB Problem in U.S. Freshwaters and Inland Waters *Joint Subcommittee on Ocean Science and Technology*	7
Chapter 3	Research on Freshwater HABs in the United States *Joint Subcommittee on Ocean Science and Technology*	31
Chapter 4	Plan for Improving Federal Response to Freshwater HABs: Research and Infrastructure Priorities *Joint Subcommittee on Ocean Science and Technology*	53
Chapter 5	Steps to Improve Coordination and Communication for Freshwater HAB Research and Response *Joint Subcommittee on Ocean Science and Technology*	77
References		83
Appendix I	Federal Freshwater/ Inland HAB Research and Response Programs *Joint Subcommittee on Ocean Science and Technology*	95

Appendix II	Other National Programs *Joint Subcommittee on Ocean Science and Technology*	**105**
Appendix III	State and Local Initiatives *Joint Subcommittee on Ocean Science and Technology*	**107**
Appendix IV	International Efforts *Joint Subcommittee on Ocean Science and Technology*	**113**
Index		**117**

PREFACE*

Freshwater harmful algal blooms (HABs) are comprised of algae that either create health hazards for humans or animals through the production of toxins or bioactive compounds or that cause deterioration of water quality through the build-up of high biomass, which degrades aesthetic, ecological, and recreational values. While freshwater HABs occur naturally, human actions that disturb ecosystems in the form of increased nutrient loadings and pollution, modified hydrology, and introduced species have been linked to the increased occurrence of some freshwater HABs. The majority of the freshwater HAB problems reported in the United States and worldwide are due to one group of algae, the cyanobacteria HABs (CyanoHABs), but other groups of algae can also be harmful. CyanoHABs are not a new phenomenon, but the frequency and geographic distribution of documented CyanoHABs seem to have dramatically increased in recent decades in the United States and globally. The issue of freshwater HABs has received more attention outside the United States in the past, but, as most U.S. states now experience freshwater HABs, the issue is of growing national concern.

---

* This is an edited, reformatted and augmented version of a Joint Subcommittee on Ocean Science and Technology publication dated 2008.

CyanoHAB scum along shoreline on Lake of the Woods. Photo: Hedy Kling (Canada)

Freshwater HAB toxins can have a broad range of negative impacts on humans, animals, and aquatic ecosystems. Many cyanobacteria can produce neurotoxic, hepatotoxic, dermatotoxic, or other bioactive compounds, and blooms of toxigenic cyanobacteria pose a particular threat if they occur in drinking water sources. The World Health Organization has issued an advisory limit for drinking water of 1 $\mu g\ L^{-1}$ for microcystin-LR, one of the most common cyanobacterial toxins, but no Federal regulations exist in the United States. The U.S. Environmental Protection Agency (EPA) has listed selected cyanobacteria and their associated toxins on the Contaminant Candidate List. EPA will use this process to determine whether it should regulate these contaminants in the future.

Adverse ecosystem impacts stemming from nontoxic high biomass blooms of cyanobacteria and other freshwater HABs have been well documented. High biomass blooms can cause low oxygen events that kill fish and bottom dwelling organisms. In addition, dense blooms block sunlight penetration, preventing the growth of other algae and disrupting food webs. Some non-cyanobacterial freshwater HABs also produce toxins that can kill fish. The most problematic of these is *Prymnesium parvum*, also called "golden algae," which has caused fish kills in Texas annually since 2001 and has been documented in at least nine other states. The human health and ecological impacts of freshwater HAB outbreaks can have serious economic and sociocultural costs, but these impacts have not been well quantified in the United States.

In 2004, Congress reauthorized the Harmful Algal Bloom and Hypoxia Research and Control Act of 1998 with the Harmful Algal Bloom and Hypoxia Amendments Act (HABHRCA 2004). The 2004 legislation required the generation of five reports (see Box 1.1), including this *Scientific Assessment of Freshwater Harmful Algal Blooms*. HABHRCA 2004 stipulates that this report 1) examine the causes, consequences, and economic costs of freshwater HABs, 2) establish priorities and guidelines for a research program on freshwater HABs, and 3) make recommendations to improve coordination among Federal agencies with respect to research on HABs in freshwater environments.

This report is divided into five chapters: Chapter 1 provides the legislative background and process for developing the report, Chapter 2 describes the problem of freshwater and inland HABs in the United States, Chapter 3 outlines the current Federal efforts in freshwater and inland HAB research and response, Chapter 4 discusses the future research priorities, and Chapter 5 delineates opportunities for coordination to advance research efforts. The document is based, in large part, on the proceedings (Hudnell 2008) of the International Symposium on Cyanobacterial Harmful Algal Blooms, a meeting convened by EPA and sponsored by a variety of Federal agencies, to describe current scientific knowledge and identify priorities for future research on CyanoHABs.

STATE OF THE SCIENCE IN THE UNITED STATES: PRIORITIES FOR ADVANCEMENT

In the United States, progress to date on freshwater HAB research and response has been made mostly through research at the individual project level with larger Federal research and response efforts concentrated on the Great Lakes region. Most efforts have been directed toward monitoring of waters used for drinking or recreation and improvements in drinking water treatment for toxins; Chapter 3 provides an overview of these and other Federal activities and advancements.

This report offers a plan for coordinating the important research that is currently ongoing in the United States and for guiding future research directions for Federal programs as well as for state, local, private, and academic institutions in order to maximize advancements. To this end, the Interagency Working Group on Harmful Algal Blooms, Hypoxia, and Human Health (IWG-4H) identifies seven priorities, all of equal weight, for freshwater HAB research and response. These priorities represent research areas where there is the greatest potential for progress in freshwater HAB research. This report does not attempt to assess the relative importance of freshwater HAB research compared to other research areas or other priorities for Federal or state investment.

Priority: Improve Methods for Detecting HAB Cells and Toxins

Tools and methods to reliably detect HAB cells in the field and HAB toxins in a variety of matrices (dissolved in water, inside algal cells, and in animal tissues) will be important for progress in all priority research areas. Identifying the occurrence of algal blooms and their toxins through reliable detection methods will help protect human and ecosystem health. Moreover, the ability to quantify HAB cells and toxins will advance the exploration of human and ecological health effects and the development of effective management strategies. Development of manual methods and automated, *in situ* sensors are both important. *In situ* sensors should be coupled with ongoing and developing observing systems.

Priority: Improve Understanding of HAB Toxin Uptake, Metabolism, and Health Effects in Humans and Animals

Some cyanobacteria-produced toxins (cyanotoxins) have been well studied; others are only partially characterized, and new cyanotoxin analogues continue to be discovered. Cyanotoxins have been a priority for research as they are a known threat to public health because of their potency and because human disease has been associated with exposure to high concentrations of these compounds. There has been less focus on other freshwater HAB toxins. Moreover, less is known about the effects of chronic, low-level toxin(s) exposure or of the effects of less-studied irritant toxins, other cyanobacteria-produced bioactive compounds, and emerging freshwater HAB toxins, such as the fish killing toxin(s) produced by *Prymnesium parvum*. Research on different toxin exposure routes and the role of metabolism in toxic responses and detoxification will improve health risk assessments.

Priority: Improve Human Health and Ecological Risk Assessments

Many goals of research on human and ecological effects overlap with those above as they require advancements in toxin research. Case studies and past animal studies demonstrate that algal toxins pose hazards to humans and ecosystems, but data does not yet exist for accurate risk assessments. For ecological effects research, broadening the scope to include entire food webs and longer time periods would offer significant improvement.

Priority: Improve knowledge of Bloom Occurrence Through Better Monitoring

Monitoring for cells and toxins will advance knowledge of bloom occurrence, which is important to understand the prevalence and severity of freshwater HABs and to prevent or reduce impacts on humans and ecosystems. An ideal monitoring system would allow for real-time, highly automated, accurate bloom detection that in the short run could provide early warning of impending toxic events and will lead to better predictive capability in the long run. Tools to advance monitoring are a focus in multiple priority research areas and include citizen monitoring networks; databases; diagnostic tools for monitoring exposure and effects; and easier, cheaper, faster, and more accurate methods for detecting cells and toxins.

Fish Kill associated with a bloom of the golden alga, *Prymnesium parvum*, on Lake Whitney in Texas Photo: Joan Glass, TPWD

Priority: Improve Bloom Prediction

Models that predict the occurrence and toxicity of HABs, bloom transport, and the fate of toxins are the goal of much HAB research. Developing predictive models requires a thorough understanding of the interaction of HAB organisms and their environment. By identifying critical factors regulating bloom dynamics, predictive models will also help refine management strategies to reduce or prevent HABs.

Priority: Develop HAB Prevention and Control Methods

Bloom prevention and control are the ultimate management goals, and real world examples have demonstrated that these management strategies can be very successful. Prevention and control are not always feasible, however, and can be expensive. Control strategies carry additional challenges due to limits of effectiveness, potential environmental impacts, and negative public perceptions. Research to overcome challenges include developing strategies, such as nutrient reduction or methods of bloom control or toxin removal, that are based on a thorough scientific understanding of bloom dynamics and involve stakeholder input so that any negative impacts to the economy and environment are minimized. Cost-benefit analysis will also be useful for evaluating the overall value of management approaches that implement prevention and control strategies.

Priority: Improve HAB Research and Response Infrastructure

Different types of infrastructure support the ability to monitor, predict, mitigate, control, and prevent HABs. Improvements in freshwater HAB infrastructure could support state-of-the-art HAB monitoring and detection and lead to more accurate risk assessments as well as short- and long-term HAB predictions. Currently, the most important infrastructure goal for achieving research priorities is to improve the availability of toxin standards and other reference materials needed to conduct research and monitoring. In addition, an event response program for freshwater HABs included in or analogous to those for marine HABs would be beneficial. Coordination of existing programs and resources will improve efficiency of freshwater HAB research, response, and management.

NEXT STEPS: IMPROVING FEDERAL COORDINATION AND ADVANCING RESEARCH

The priorities and goals presented in this report offer direction and focus to advance freshwater HAB research and response. Federal coordination will streamline these research efforts by fostering collaboration and minimizing unnecessary duplication. The IWG-4H, as the body fulfilling the role of the Interagency Task Force on HABs and Hypoxia, provides Federal coordination for HAB research and response. The IWG-4H may also cultivate Federal coordination through interaction with the National HAB Committee, which represents the HAB research and management communities. Coordination at the international level, such as through participation in national and international meetings and the global cyanobacteria network, CYANONET (http://www.cyanonet.org), will also be important for accelerating advancements and avoiding duplication because other countries have considerable experience addressing the freshwater HAB problem.

HABHRCA 2004 calls for a "competitive, peer-reviewed, merit-based interagency research program as part of the Ecology and Oceanography of Harmful Algal Blooms (ECOHAB) [program], to better understand the causes, characteristics, and impacts of harmful algal blooms in freshwater locations..." The interagency ECOHAB Program—comprised of the EPA, National Oceanic and Atmospheric Administration, National Science Foundation, National Aeronautics and Space Administration, and Office of Naval Research—currently meets this requirement. However, because of the mandates of some agencies involved, the freshwater focus has been on the Great Lakes and upper reaches of estuaries.

LIST OF ACRONYMS

AOAC	Association of Analytical Communities
AwwaRF	American Water Works Association Research Foundation
BOR	U.S. Bureau of Reclamation
CCFHR	NCCOS Center for Coastal Fisheries and Habitat Research, NOAA
CCL	Contaminant Candidate List
CCMA	NCCOS Center for Coastal Monitoring and Assessment, NOAA
CDC	Centers for Disease Control and Prevention
CSCOR	NCCOS Center for Sponsored Coastal Ocean Research, NOAA
CWA	Clean Water Act
CyanoHAB	Cyanobacteria harmful algal bloom
ECOHAB	Ecology and Oceanography of Harmful Algal Blooms
ELISA	Enzyme-linked immunosorbent assay
EPA	U.S. Environmental Protection Agency
FDA	U.S. Food and Drug Administration
GEOHAB	Global Ecology and Oceanography of Harmful Algal Blooms Program
GLERL	Great Lakes Environmental Research Laboratory
GOOS	Global Ocean Observing System
GWRC	Global Water Research Coalition
HAB	Harmful algal bloom
HABHRCA	Harmful Algal Bloom and Hypoxia Research and Control Act
HABISS	Harmful Algal Bloom-related Illness Surveillance System

HARRNESS	Harmful Algal Research and Response: A National Environmental Science Strategy 2005–2015
IOOS	Integrated Ocean Observing System
ISOC-HAB	International Symposium on Cyanobacterial Harmful Algal Blooms
IWG-4H	Interagency Working Group on Harmful Algal Blooms, Hypoxia, and Human Health
JSOST	Joint Subcommittee on Ocean and Science Technology
MERHAB	Monitoring and Event Response for Harmful Algal Blooms Program, NOAA
MD DNR	Maryland Department of Natural Resources
NAWQA	National Water Quality Assessment Program
NCEA	National Center for Environmental Assessment, EPA
NCEH	National Center on Environmental Health, CDC
NCER	National Center for Environmental Research, EPA
NERL	National Exposure Research Laboratory, EPA
NHC	National HAB Committee
NHEERL	National Health and Environmental Effects Research Laboratory, EPA
NIEHS	National Institute of Environmental Health Sciences
NOAA	National Oceanic and Atmospheric Administration
NSF	National Science Foundation
OHHI	Oceans and Human Health Initiative, NOAA
RDDTT	Research Development Demonstration and Technology Transfer
SeaWiFs	Sea-viewing Wide Field-of-view Sensor
SBIR	Small Business Innovation Research
SDWA	Safe Drinking Water Act
SUNY-ESF	State University of New York College of Environmental Science and Forestry
TPWD	Texas Parks and Wildlife Department
UCM	Unregulated Contaminant Monitoring
UNCW	University of North Carolina Wilmington
UNESCO	United Nations Educational, Scientific, and Cultural Organization
USACE	U.S. Army Corps of Engineers
USAMRIID	U.S. Army Medical Research Institute of Infectious Diseases
USFWS	U.S. Fish and Wildlife Service

USGS	U.S. Geological Survey
USDA	U.S. Department of Agriculture
UTEX	University of Texas
WHO	World Health Organization

In: Marine and Freshwater Harmful Algal Blooms ISBN: 978-1-60741-838-2
Editor: Peter E. Williams © 2010 Nova Science Publishers, Inc.

Chapter 1

LEGISLATIVE BACKGROUND AND REPORT PROCESS

Joint Subcommittee on Ocean Science and Technology

1.1. LEGISLATIVE BACKGROUND

The Harmful Algal Bloom and Hypoxia Amendments Act of 2004 (Public Law 108-456) (HABHRCA 2004) reauthorized the original Harmful Algal Bloom and Hypoxia Research and Control Act of 1998 (Public Law 105-383), reconstituted the Interagency Task Force on HABs and Hypoxia, and stipulated the generation of five reports (Box 1.1) to assess and recommend research programs on harmful algal blooms (HABs) and hypoxia in U.S. waters. The Interagency Task Force on HABs and Hypoxia was incorporated into the Interagency Working Group on Harmful Algal Blooms, Hypoxia, and Human Health (IWG-4H, see Box 1.2) of the Joint Subcommittee on Ocean Science and Technology (JSOST). The IWG-4H was tasked with implementing the requirements of both HABHRCA 2004 and the Interagency Oceans and Human Health Research Program established by the Oceans and Human Health Act of 2004 (see Box 1.3).

HABHRCA 2004 requires a Scientific Assessment of Freshwater Harmful Algal Blooms and stipulates that this report: 1) examine the causes, consequences, and economic costs of freshwater HABs, 2) establish priorities and guidelines for a research program on freshwater HABs, and 3) make recommendations to improve coordination among Federal agencies with respect to research on HABs in freshwater environments. For all U.S. freshwaters, this report complements and expands HAB-related priorities identified under the themes of "Improving Ecosystem Health" and "Enhancing Human Health" in *Charting the Course for Ocean Science in the United States for the Next Decade: An Ocean Research Priorities Plan and Implementation Strategy*, prepared by the JSOST pursuant to the *U.S. Ocean Action Plan.*

BOX 1.1. HABHRCA 2004 CALLS FOR THE FOLLOWING FIVE REPORTS OR ASSESSMENTS

- National Assessment of Efforts to Predict and Respond to Harmful Algal Blooms in U.S. Waters (Prediction and Response Report)
- National Scientific Research, Development, Demonstration, and Technology Transfer Plan for Reducing HAB Impacts (RDDTT Plan)
- Scientific Assessment of Freshwater Harmful Algal Blooms
- Scientific Assessment of Marine Harmful Algal Blooms
- Scientific Assessment of Hypoxia

Freshwater HABs are a growing problem globally, and several countries have guidelines for algal toxins in drinking and/or recreational waters (see Codd et al. 2005, NHMRC 2004, NHMRC 2006).

The U.S. Environmental Protection Agency (EPA) is authorized to protect human health and the natural environment from contaminants in water through the Safe Drinking Water Act (SDWA) and the Clean Water Act (CWA). The EPA listed select cyanobacteria and cyanobacterial toxins in 1998 on the first drinking water Contaminant Candidate List (CCL) and in 2005 on the second list, thereby making cyanobacteria and their toxins a priority for possible regulatory determination and requiring further information on occurrence, persistence, health risks, and remediation techniques. EPA has recently prepared draft *Toxicological Reviews of Cyanobacterial Toxins: Anatoxin-a, Cylindrospermopsin,* and *Microcystins (LR, RR, YR and LA)* (http://cfpub.epa.gov/ncea/cfm/recordisplay.cfm?deid=161263) in order to compile and evaluate the available data regarding toxicity of these algal toxins to aid in the regulatory decision-making process under the CCL.

Algal mats in Lake Ontario. *Photo: Greg Boyer, SUNY ESF*

BOX 1.2. INTERAGENCY WORKING GROUP ON HABS, HYPOXIA, AND HUMAN HEALTH (SPECIFIED IN HABHRCA AS INTERAGENCY TASK FORCE; MODIFIED BY JSOST)

- Department of Commerce, Co-chair
- Department of Health & Human Services, Co-chair
- Environmental Protection Agency
- National Science Foundation
- National Aeronautics & Space Administration
- Department of the Navy
- Department of Agriculture
- Department of Interior
- Food & Drug Administration
- Office of Science & Technology Policy
- Council on Environmental Quality

1.2. REPORT PROCESS

This report, the *Scientific Assessment of Freshwater Harmful Algal Blooms*, will 1) assess the problem of freshwater and inland HABs in the United States, including algal species that bloom in saline inland waterbodies, such as the Salton Sea, 2) detail current Federal efforts in freshwater and inland HAB research and response, and 3) identify future research priorities and opportunities for coordination to advance these efforts.

Cyanobacteria HABs (CyanoHABs) are the most prevalent and problematic HABs in freshwaters worldwide. Much research has been conducted internationally on the management of CyanoHABs over the last 20 years (summarized in WHO 1999, Falconer 2005, Huisman et al. 2005, Hudnell 2008), and numerous management strategies and global coordination efforts have been initiated (e.g. WHO 1999, WHO 2000, GWRC 2004, Codd et al. 2005). The triennial International Conference on Toxic Cyanobacteria series began in 1980 to maintain awareness of advances in CyanoHAB research. The international experiences and the existing global coordination efforts for CyanoHABs were important considerations when developing this report.

In order to describe current scientific knowledge and directions for further research on the growing problem of CyanoHABs, the EPA established an interagency committee to organize an International Symposium on Cyanobacterial Harmful Algal Blooms (ISOC-HAB), sponsored by U.S. Army Corps of Engineers (USACE), Centers for Disease Control and Prevention (CDC), EPA, U.S. Food and Drug Administration (FDA), National Oceanic and Atmospheric Administration (NOAA), National Institutes of Health (NIH), U.S. Department of Agriculture (USDA), U.S. Geological Survey (USGS), and University of North Carolina at Wilmington (UNCW). Attendees included U.S. researchers and managers as well as international experts from countries already facing CyanoHAB problems. The following major topic areas were addressed in symposium workgroups: 1) occurrence of cyanobacteria and toxins, 2) causes, prevention, and mitigation, 3) cyanotoxin characteristics, 4) human health and ecological effects, 5) analytical methods for identifying and quantifying cyanobacteria and toxins, and 6) risk assessment. The research papers, poster session abstracts, and workgroup reports, including priority research topics for each of the focus areas, are being published as a peer-reviewed monograph, *Cyanobacterial Harmful Algal Blooms: State of the Science and Research Needs* (Hudnell 2008). These proceedings, which include reviews and summaries of research conducted internationally on CyanoHABs over the last two decades, contributed in large part to the scientific assessment (see Chapter 2) and research priorities (see Chapter 4) in this report.

Although CyanoHABs are the most prevalent freshwater HAB, other HABs have caused problems in freshwaters and other inland waters. The scientific literature was the primary source of information used to assess the problem and state-of-the-science on non-cyanobacterial freshwater HABs for the purpose of this report.

Another important component of this report is to detail current efforts pursued by Federal, state, and local governments to advance freshwater HAB research in the areas described above (see Chapter 3). Federal agencies were individually queried about programs or projects relevant to freshwater HABs, and details about state and local activities were primarily collected from information provided on Federal, state, and local government websites.

BOX 1.3. OCEANS AND HUMAN HEALTH (OHH) ACT 2004 (PUBLIC LAW 108-447)

The OHH Act requires the National Science and Technology Council to establish an Interagency Oceans and Human Health Research Program to improve understanding of the role of the oceans in human health and establishes the NOAA Oceans and Human Health Initiative as part of this interagency program. The JSOST IWG-4H, in addition to serving as the Interagency Taskforce on Harmful Algal Blooms and Hypoxia as called for in HABHRCA, was charged with the responsibility for coordinating the interagency OHH program and producing both the HAB-related and OHH-related reports to Congress. HABs are included as part of the OHH program scope, but the OHH Act specifically states that "nothing in this subsection is intended to duplicate or supersede the activities of the Inter-Agency Task Force on Harmful Algal Blooms and Hypoxia." The IWG-4H has prepared a 10-year Interagency OHH Implementation Plan (Sandifer et al. 2007), which was called for by the OHH Act. Coordination with HABHRCA activities is provided through the IWG-4H since it has responsibilities for both OHH and HABs and hypoxia.

This report offers a plan for coordinating the research that is currently ongoing in the United States and for guiding Federal and other research and grant programs in developing new directions for research that may lead to improved management strategies.

In: Marine and Freshwater Harmful Algal Blooms ISBN: 978-1-60741-838-2
Editor: Peter E. Williams   © 2010 Nova Science Publishers, Inc.

Chapter 2

THE HAB PROBLEM IN U.S. FRESHWATERS AND INLAND WATERS

Joint Subcommittee on Ocean Science and Technology

2.1. GENERAL OVERVIEW

Freshwater HAB taxa are comprised of algae that either create health hazards for humans and/or animals or cause deterioration of water quality or aesthetic/ recreational values (see Table 1). By far the most problematic and researched group of freshwater HABs are the cyanobacteria (formerly called blue-green algae). CyanoHABs form high biomass blooms and/or produce toxins, as well as taste-and-odor compounds, and have caused human illness, animal mortalities, and adverse ecosystem and economic impacts in the United States and worldwide, as well as human mortalities in some other nations (Hudnell 2008). Other freshwater HAB taxa cause harm either by producing toxins that kill fish and other aquatic organisms or by forming high biomass blooms that can cause hypoxia (low dissolved oxygen) and degrade water quality in other ways. Harmful algae that cause harm in saline inland waterbodies and upper reaches of estuaries are also covered in this report.

Problems related to freshwater/inland HABs are widespread and have become more prevalent in recent decades. For example, toxic cyanobacterial outbreaks seem to be expanding and occurring more frequently in U.S. waters and globally, a trend reflected in the increasing number of published studies and reports (see Figure 1, Carmichael 2008). There have also been increasing reports of harmful species emerging in areas that did not have problems in the recent past, such as the cyanobacteria, *Cylindrospermopsis*, in eutrophic lakes in Florida (Chapman and Schelske 1997, Figure 2) and in Chesapeake Bay tributaries in Maryland. In addition, cyanobacteria-produced toxins and taste-and-odor compounds are becoming more of a problem in drinking water reservoirs (Izaguirre 2008), and off-flavor compounds have become particularly problematic in aquaculture operations (Tucker 2000). Non-cyanobacterial inland HAB events are increasing in some locations as well. For example, the golden alga, *Prymnesium parvum*, which became a problem in Texas in the 1980's, has caused fish kills annually since 2001 and has now spread up the Pecos River into New Mexico and down the Red River into Lake Texoma and Oklahoma.

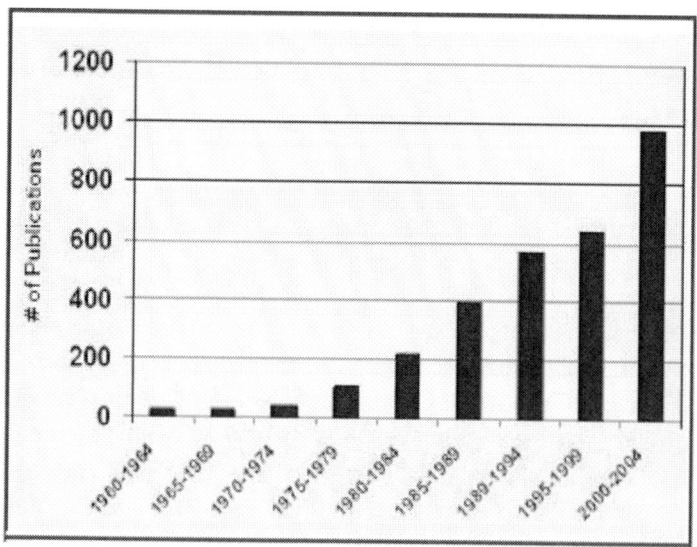

Figure 1. Number of published CyanoHAB scientific studies and reports logged worldwide (Carmichael 2008)

Algae proliferate when multiple interacting biological, chemical, and physical factors act synergistically to create an optimal growth environment. Those optimal growth conditions vary among algal species, even within the freshwater HAB taxa. Understanding the causative factors and how they interact to foster bloom formation and maintenance is critical for developing effective management strategies. While HABs occur naturally, human actions that disturb ecosystems have been linked to the rise in some freshwater HAB outbreaks (Paerl 2008). The freshwater HAB taxa (see Table 1), the potential causes of their proliferation, where they occur, and their impacts are discussed in the following sections.

Table 1. Harmful or Nuisance Inland Algae, the Types of Toxins They Produce, the Direct Impacts of Blooms, and the Geographic Locations Where Problems Have Been Documented in the United States

Inland HAB Taxa (Specific organism/s of concern)	Toxins	Adverse Impacts	Impacted Areas of the United States
Cyanobacteria (many species, see Table 2 for toxigenic genera)	Hepatotoxins, neurotoxins, cytotoxins, dermatotoxins, respiratory and olfactory irritant toxins (See Table 2)	Human and animal health impacts (see Table 2); water discoloration, unsightly and foul-smelling scums; hypoxia from high biomass blooms; taste-and-odor problems in drinking water and in farm raised fish	Majority of U Figure 3)
Haptophytes (e.g., *Prymnesium parvum*, *Chrysochromulina polylepis*)	Ichthyotoxins	Mortalities of fish and other gill breathing species	*P. parvum*: Major fish kills in TX & NM. Also confirmed in AK, AL, CO, GA, NC, SC, WY, OK. Suspected in FL and NE
Chlorophytes Microalgae *Volvox*, (e.g., *Pandorina*)	--	Discolored water, localized hypoxia	Small eutrophic ponds
Macroalgae (e.g., *Cladophora*)	--	Unsightly and foul-smelling mats, localized hypoxia, clogged water intakes	Great Lakes, FL inland lakes
Euglenophytes (*Euglena sanguinea*)	Ichthyotoxin (Zimba et al. 2004)	Discolored water, mortalities of fish	NC (Zimba et al. 2004)
Raphidophytes *Chattonella* (marine but blooms in inland saline waters)	Ichthyotoxins	Fish Kills	Occurrence and potential cause of fish Sea (Reifel et al. 2002, SSERG 2001)

Table 1 (Continued)

Inland HAB Taxa (Specific organism/s of concern)	Toxins	Adverse Impacts	Impacted Areas of the United States
Dinoflagellates (e.g., *Peridinium polonicum* [syn. *Peridiniopsis polonicum*, *Glenodinium Gymnodinium*])	Icht-hyotoxins (Oshima et al. 1989)	Mortalities of fish	Discolored water in OK (Nolen et al. 1989). Fish kills in Japan and Spain (Oshima et al. 1989, Roset et al. 2002)
Cryptophytes	--	High biomass blooms can cause discolored water, localized hypoxia	Not normally thought of as nuisance taxa
Diatom (*Didymosphenia geminata*)	--	Produces large amounts of extracellular stalk material resulting in ecosystem and economic impacts	Nuisance blooms found in streams and rivers

2.2. CYANOBACTERIA

Cyanobacteria are the major harmful algal group in freshwater environments and are recognized as a rapidly expanding global problem that threatens human and ecosystem health (Carmichael 2008). CyanoHABs can manifest as visual discolorations in waterbodies and, at times, as surface scums that appear as paint-like slicks or clotted mats. The color of surface scums is most often light green to dark, brownish green but can also be red to reddish brown. Harmful effects of cyanobacteria can occur even when visible signs of a bloom are absent. CyanoHABs have been linked to animal deaths and human illness all over the world (Carmichael 2008, Stewart 2008, Box 2.1). Efforts have been underway internationally for almost two decades to develop guidelines for managing human health risks associated with cyanobacteria and their toxins in drinking and recreational waters (e.g., NRA 1990, WHO 1998, WHO 2000, WHO 2003, GWRC 2004, Codd et al. 2005).

Cyanobacteria, through their long history of evolution, have developed unique adaptive capabilities that allow them to take advantage of variable environmental conditions. Consequently, they can live and proliferate in many different environments. CyanoHABs are considered one of the most obvious indicators of nutrient over-enrichment (Paerl and Fulton 2006), and successful watershed management to reduce excessive nutrient loadings has been shown to decrease CyanoHAB occurrence in some areas (Box 2.2, Edmondson and Lehman 1981). In addition to nitrogen and phosphorus enrichment, hydrologic modifications that alter water flushing rates, food web changes (such as the removal of grazers), and introduction of toxins and pollutants are all important factors influencing CyanoHAB formation (Paerl 2008). Better understanding of the interactive role of these and other factors (such as nutrient ratios, organic matter availability, climate change, temperature, salinity, light attenuation, freshwater discharge, and water column stability) in bloom formation will lead to new approaches for prediction, response, and prevention (Perovich et al. 2008). A better understanding of which environmental parameters control toxin production by cyanobacteria is also needed.

In the United States, CyanoHABs have been documented in at least 35 states (Figure 3). Recent surveys in Nebraska, New York, New Hampshire, and Florida indicate that CyanoHAB abundance has increased in these states in recent years (Fristachi et al. 2008), and at least 18 states now have some type of CyanoHAB research or response program (see Appendix III). Many cyanobacterial species have been shown to produce neurotoxic, hepatotoxic, and dermatotoxic compounds (Fristachi et al. 2008, Table 2), and toxicity is estimated to occur in a majority of all cyanobacterial blooms in the United States (Haney and Ikawa 2000, Boyer et al. 2004, Graham et al. 2004) and worldwide (WHO 1999).

Oscillatoria rubescens bloom in Georgia. *Photo: Michael Smith, Identification: Steve Morton, NOAA*

Table 2. Cyanobacterial Toxins, the Freshwater Taxa That Produce Them (Paerl 2001, Fristachi et al. 2008), and Human Health Effects (Harrness 2005, Falconer 2005). Not All Species/Clones of Genera Listed Produce Toxins. Environmental Conditions May Also Influence Toxin Production

Toxin	Genera	Short Term Health Effects	Long Term Effects
Microcystins	*Anabaena, Aphanocapsa, Hapalosphon, Microcystis, Nostoc, Oscillatoria, Planktothrix*	Gastrointestinal, liver inflammation, and hemorrhage and liver failure leading to death, pneumonia, dermatitis	Tumor promoter, liver failure leading to death
Nodularins	*Nodularia spumigena*	Similar to Microcystins	Similar to Microcystins
Saxitoxins	*Anabaena, Aphanizomenon, Cylindrospermopsis, Lyngbya*	Tingling, burning, numbness, drowsiness, Incoherent speech, respiratory paralysis leading to death	Unknown
Anatoxins	*Anabaena, Aphanizomenon, Oscillatoria, Planktothrix*	Tingling, burning, numbness, drowsiness, incoherent speech, respiratory paralysis leading to death	Cardiac arrhythmia leading to death
Cylindrospermopsin	*Aphanizomenon, Cylindrospermopsis, Umezakia*	Gastrointestinal, liver inflammation and hemorrhage, pneumonia, dermatitis	Malaise, failure leading to death
Lipopolysaccharide	*Aphanizomenon, Oscillatoria*	Gastrointestinal, dermatitis	Unknown
Lyngbyatoxins	*Lyngbya*	Dermatitis	Skin tumors 1990), unknown
BMAA	*Anabaena, Cylindrospermopsin, Microcystis, Nostoc, Planktothrix*		Potential link to neurodegenerative diseases

> ## BOX 2.1. EXAMPLES OF KNOWN HUMAN IILLNESS OR DEATH ASSOCIATED WITH CYANOHABS WORLDWIDE (MODIFIED FROM WHO 2003)
>
> **Australia**
>
> *1979:* Serious illness and hospitalization of 141 people associated with toxic bloom in drinking water reservoir, which had been treated with copper sulfate (Falconer 1993).
> *1995:* Human illness (gastrointestinal) associated with recreational water contact in waters with cyanobacteria (Pilotto et al. 1997).
>
> **Brazil**
>
> *1988:* Death of 88 and illness of around 2,000 people associated with toxic cyanobacteria in drinking water reservoir after flood (Teixera et al. 1993).
> *1996:* Death of 52 dialysis patients and illness of 64 others associated with microcystin toxins in water used for dialysis (Jochimsen et al. 1998, Carmichael et al. 2001).
>
> **Canada**
>
> *1959:* Illness (headache, muscular pains, gastrointestinal) of 13 people after recreational exposure to cyanobacterial bloom (Dillenberg and Dehnel 1960).
>
> **China**
>
> *1993:* Liver cancer incidence found higher for populations using surface waters where cyanobacteria occurred in drinking water rather than groundwater (Yu 1995), although other factors, including hepatitis and exposures to aflatoxin, may have contributed.

> **England**
>
> *1989:* Illness in soldiers training in water with cyanobacterial bloom (Turner et al. 1990).
>
> **Sweden**
>
> *1994:* Illness (gastrointestinal and muscle cramps) of 121 (out of 304) inhabitants of a village whose drinking water supply was accidentally cross-connected with cyanobacterial contaminated untreated river water (Anadotter et al. 2001).
>
> **USA**
>
> *1931:* Illness of around 8000 people whose drinking water came from tributaries of Ohio River, where a large cyanobacteria bloom had occurred (Miller and Tisdale 1931)
> *1968:* Gastrointestinal illnesses were documented to occur in association with massive blooms of cyanobacteria (cases compiled by Schwimmer and Schwimmer, 1968).
> *1974:* Chills, fever, and hypotension in 23 dialysis patients in Washington, D.C. associated with cyanobacteria in a local water source (Hindman et al. 1975).
> *2004:* Gastrointestinal illness and dermal irritation associated with recreational exposure to a CyanoHAB event in Nebraska (Walker et al. 2008).

The most serious impacts of CyanoHABs derive from their production of these potent "cyanotoxins." The majority of impacts in the United States have included taste-and-odor problems in drinking water (Izaguirre 2008) and aquaculture resources (Tucker 2000, Box 2.3), animal deaths (Stewart 2008, Boxes 2.4-2.5), and reduced recreational opportunities (Walker et al. 2008, Boxes 2.4-2.10). Other impacts in the United States include human illnesses associated with the presence of large blooms in recreational waters (Walker et al. 2008) or drinking water sources (Box 2.1).

2.2.1. Impacts of CyanoHABs on Human Health

Cyanotoxins, especially hepatotoxins, may be a health threat to humans and other animals exposed via contaminated water. Cyanotoxins have been linked to human deaths in Brazil (Box 2.1), but to date, no human deaths in the United States have been unquestionably tied to cyanotoxins. Selected cyanobacteria and their toxins are included on the EPA's CCL (http://www.epa.gov/safewater/ccl/ccl2.html). EPA is assessing their occurrence in drinking water and their health effects in order to determine if actions regarding drinking water guidance, health advisories, or regulations are necessary.

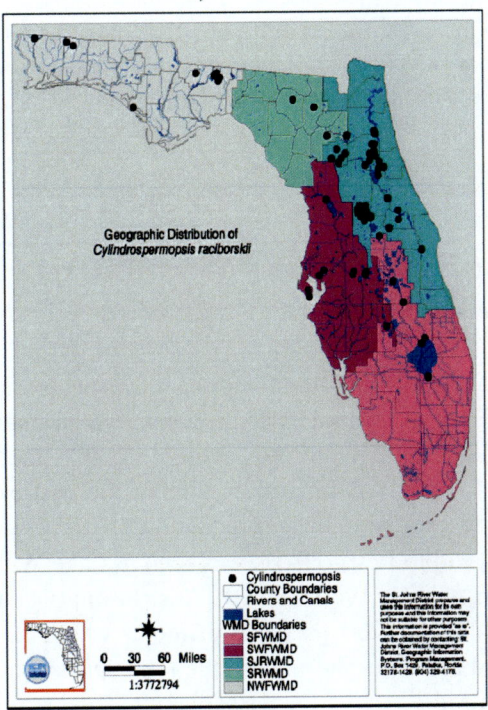

Figure 2. Distribution of the CyanoHAB, *Cylindrospermopsis raciborskii,* in Florida (Williams 2001, Fristachi et al. 2008).
C. raciborskii, which produces potent hepatotoxins (Table 2), was originally found only in tropical areas but has recently spread to cooler regions.

Figure 3. Approximate location of some freshwater CyanoHAB outbreaks documented in the United States from 1883 to 2005 (modified from Carmichael 2008).
Dots are geographically general representations of outbreaks and may represent events occurring in more than one waterbody within a state. The map is not a complete representation of all CyanoHAB outbreaks.

BOX 2.2. CASE STUDY: LAKE WASHINGTON NUTRIENT REDUCTION LED TO PREVENTION OF CYANOHABS

Prevention is a crucial, long-term approach for successful management of freshwater HABs. For CyanoHABs, preventative strategies commonly target excess nutrients and hydrology, which are two factors known to strongly influence CyanoHAB formation. One of the best examples in the United States of how targeted nutrient reduction can reverse the effects of eutrophication and lead to fewer HABs was seen in Lake Washington (near Seattle, Washington) in the 1960's and 1970's. From 1941 to 1963, Lake Washington received secondary sewage effluent, which introduced large amounts of nitrogen and phosphorus into the lake. Disproportionately high phosphorus concentrations led to a phytoplankton community dominated by cyanobacteria (almost 100% in the summer months) from 1955 to 1973. Sewage was diverted from the lake beginning in 1963, and the amount of untreated effluent entering the lake was reduced to near zero by 1968. Water quality improvements rapidly followed, and cyanobacteria decreased and have been a relatively insignificant member of the phytoplankton community since 1976 (Edmonson and Lehman 1981).

Drinking water contaminated with cyanobacteria can have taste-and-odor problems due to nontoxic compounds, which may sometimes prevent ingestion of toxin-contaminated water by alerting utilities to a CyanoHAB problem. However, toxic cyanobacteria can also occur without associated taste or odor problems. The extent of the threat to humans via drinking water is not completely clear, but the first nationwide survey done on 45 utility waters (i.e., source waters, treatment and plant intakes, and plant effluents) in the United States and Canada between 1996 and 1998 found 80% of the 677 samples to contain detectable levels of microcystins (Carmichael 2001). About 4% exceeded the World Health Organization's (WHO) advisory limit of 1 µg L^{-1} for microcystin-LR in drinking water, but only two of the 4% were "finished water" samples, indicating the standard water treatment was relatively effective at removing cyanotoxins. More recently, samples of untreated source water taken from Lake Erie (Boyer 2008) and of finished water in Florida (Burns 2008) have exceeded the WHO's advisory limit.

BOX 2.3. OFF-FLAVOR IN CATFISH FARMING CAUSES SIGNIFICANT ECONOMIC LOSSES

Catfish farming is the largest aquaculture industry in the United States, with an annual production valued at $450 million. Off-flavor in catfish delays harvesting and increases production costs for catfish farmers and is the second largest cause of economic losses to the industry. The cyanobacteria-produced compounds geosmin (earthy tasting) and 2-methylisoborneol (MIB) (musty tasting) account for about 80% of these losses. In a thorough review of off-flavor issues in aquaculture, Tucker (2000) estimated that economic losses ranged from 3 to 17% of the growers' selling price, or as high as $60 million at 1998 prices. Additional costs to the industry include potential reductions in market value due to decreased demand as a result of inconsistency in product quality. Direct mortalities of catfish from microcystins have also been documented (Zimba et al. 2001).

The presence of high levels of cyanotoxins in drinking water can cause gastrointestinal complications, liver damage, neurological symptoms, and potentially, although rare, even death (Falconer 2008, Box 2.1). The public health impacts of chronic, low-level exposures are unknown (Backer 2002, Hilborn et al. 2008). Recent research has hypothesized a possible link between a cyanobacterial-produced amino acid (β-methylamino-L-alanine or BMAA) and neurological diseases, such as Alzheimer's (Cox et al. 2003, Murch et al. 2004, Cox et al. 2005), but much work needs to be done to verify this link.

Recreational exposure to toxic CyanoHABs via direct skin contact, inhalation, or inadvertent ingestion of water can cause rashes, allergies, and gastrointestinal problems for people working or recreating on the water (WHO 2003). The long-term effects of such exposures or the effects of inhalation of toxins are not well known either. Pilotto et al. (1997) discussed that because the adverse effects of cyanotoxins restrict their use in experimental studies, it is difficult to develop scientifically-based safety guidelines for human ingestion and inhalation.

2.2.2. Impacts of CyanoHABs on Aquatic Ecosystems and Domestic Animals

Ecosystem impacts stemming from the effects of nontoxic, high biomass CyanoHABs are well documented (Fournie et al. 2008). Low oxygen events that suffocate and kill fish and bottom-dwelling organisms are perhaps the most common adverse impact of high biomass blooms.

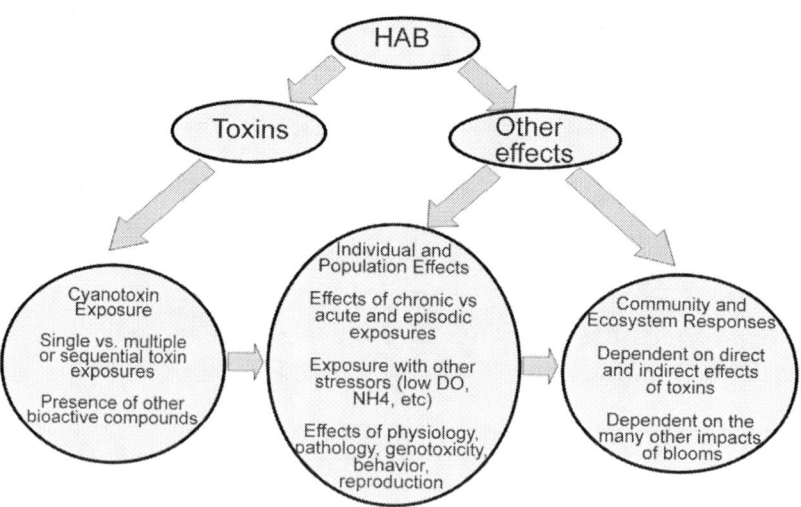

Figure 4. Diagram illustrating the complexity of CyanoHAB effects on ecosystems (Fournie et al. 2008).

In addition, high biomass blooms can block sunlight from penetrating into the water column, thereby preventing growth of beneficial algae. Food web crashes can also result due to the unpalatability and low food quality of many cyanobacteria, which can result in the starvation of consumers and their predators. The toxicity associated with many CyanoHABs can exacerbate the nature of these impacts on aquatic ecosystems (Figure 4), but the importance of toxicity relative to the stressors described above is unclear (Havens 2008, Ibelings et al. 2008). Cyanotoxins can accumulate in the primary consumers (Prepas et al. 1997) and potentially be transferred up the food web. Cyanotoxins have been implicated as the cause of mass mortalities of fish (Peñaloza et al. 1990, Tencalla et al. 1994, Rodger et al. 1994) and birds (Koeing 2006), and have also been tied to the death of pets (Boxes 2.5, Walker et al. 2008) and livestock (see Carmichael 1992, Box 2.4), which may be exposed through drinking contaminated water or licking themselves after bodily exposure. Furthermore, ungrazed cyanobacterial biomass that accumulates as clotted mats poses a particular threat to dogs, who may lick or eat the toxic mats. For this reason, dog deaths have emerged as an unfortunate early warning in many cases that a toxic CyanoHAB is occurring (Walker et al. 2008). Stewart (2008) gives a thorough review of cases where cyanotoxins have been implicated in wildlife, livestock, and pet mortalities.

Box 2.4. CyanoHABs raise human health concerns in Oregon

Potentially toxigenic cyanobacteria have resulted in health advisories in 26 lakes across Oregon. Cyanotoxins have been attributed to the poisoning of twenty bighorn sheep in Owyhee Reservoir and several dog deaths in the John Day River. In 2001, a major bloom of *Anabeana flos-aquae* that produced the neurotoxin anatoxin-a was detected at Diamond Lake. The lake was closed and posted by the local county health department and U.S. Forest Service. Since that time, several recreational lakes across Oregon have tested positive for cyanotoxins, predominantly microcystins. In response to the Diamond Lake episode, numerous stakeholders, including government, academic, and private interests, convened an interagency cyanobacteria taskforce to develop recommendations for recreational guidance. In 2004, a toxic bloom in Lake Selmac, used for potable water, caused the state to issue a health advisory and supply bottled water to consumers until a toxin level below the WHO advisory limit was achieved.

Anabaena bloom in Daly Lake, Oregon. Photo: U.S. Forest Service (Mt. Hood National Forest)

BOX 2.5. CYANOHABS IN NEW YORK AND VERMONT WATERS AND LOWER GREAT LAKES

CyanoHABs are common throughout New York state waters. A large outbreak of toxic *Microcystis* in Lake Erie in the mid-1990's raised awareness of the problem in the region. More recently, dog and waterfowl deaths have been tied to toxic blooms in Lake Champlain and Lake Neahtawanta in New York. Studies, funded in part by NOAA MERHAB, CDC, and NOAA Sea Grant, to understand occurrence in New York waters and the lower Great Lakes sampled over 1000 sites at 81 lakes during 2000-2004. These surveys revealed that nearly 60% of 2,500 lake water samples collected during the bloom season had detectable levels of microcystins with approximately 15% exceeding the WHO advisory limit for drinking water (Boyer 2008). In addition, CDC supported a more targeted public health response by funding additional sampling in response to dog deaths. Both the states of New York and Vermont participate in the Lake Champlain Basin Program (see Appendix III) which issues alerts on CyanoHABs in the lake and posts a map of the lake with existing bloom status.

Sign to warn public of CyanoHAB dangers posted in Lake Neatahwanta, NY, Photo: *Greg Boyer, SUNY ESF*

BOX 2.6. CYANOHABS IN IOWA RECREATIONAL WATERS

The Iowa Department of Natural Resources (IDNR), in conjunction with Iowa State University, began monitoring Iowa waters for cyanobacteria in 2000 and toxins in 2003. Despite low levels of measured toxin in 2003, IDNR began an intensive investigation of CyanoHABs in Carter Lake (which is located on the Iowa/Nebraska state line) in 2004, partially as a pro-active response to increasing reports of CyanoHABs in surrounding states (e.g., Walker et al. 2008). In the summer of 2004, green water and swimmers with rashes suggested a potential CyanoHAB in Carter Lake, and sampling by IDNR and Nebraska Department of Environmental Quality confirmed high levels of cyanobacteria and microcystin toxins in the Lake. The agencies used Nebraska's established toxicity guidelines of 15 parts per billion (ppb) for microcystins (Walker et al. 2008) to determine public health risk and, as a result of samples exceeding this limit, issued a health alert to warn against full-body contact with water at Carter Lake. Iowa monitors 132 lakes for cyanobacteria and microcystins, but, to date, Carter Lake has had the highest toxin of Iowa's Lakes and is the only one that has exceeded the recreational alert limit of 15 ppb.

CyanoHAB in a Nebraska Lake in 2006, Photo: Nebraska Department of Environmental Quality

BOX 2.7. CYANOHAB OCCURRENCE ALONG THE KLAMATH RIVER

In 2005 and 2006, two reservoirs, Copco and Iron Gate, along the Klamath River experienced prolonged toxic blooms of *Microcystis aeruginosa*. Toxin was also detected near tribal lands downstream, where local tribal members rely on the River for subsistence fishing. In response

> to a cyanobacteria workshop in 2005 sponsored by the state and EPA, a taskforce of county, state, Federal, and tribal authorities initiated development of statewide guidance for California on CyanoHABs. In addition, a Klamath-specific working group is meeting to administer a three-year study of the occurrence, distribution, and causes of cyanobacterial blooms in the Klamath River.

2.2.3. Economic and Sociocultural Impacts of CyanoHABs

CyanoHABs can have significant economic and sociocultural impacts due to the human health threats and their negative impact on aquaculture, recreation, and tourism. Unfortunately, these impacts have not been well quantified and documented in the United States. Impact assessments of this type have been identified as an opportunity for advancement in another HABHRCA 2004 report (Jewett et al. 2007). As an example of the potential magnitude of economic losses, overall costs in Australia have been estimated between A$180 million and A$240 million per year (Atech 2000, reviewed by Steffensen 2008), which would be equivalent to about $150-200 million in U.S. 2002 dollars.

In the United States, toxins and taste-and-odor compounds (geosmin and 2-methylisoborneaol, or MIB) result in increased treatment costs for drinking water facilities, and algal mats can interfere with reservoir operations, such as drinking water intakes and hydroelectric generation. Furthermore, off-flavor compounds are second to disease as the major cause of economic losses for catfish aquaculture and have been estimated to result in as much as $60 million in annual economic losses for the industry (at 1998 prices, which is about $72 million in 2005 dollars) (Tucker 2000, Box 2.3).

Closures of recreational waterbodies to protect human health can result in revenue losses for local communities, especially during holiday weekends or planned events. Reduced aesthetic appeal and negative perceptions of drinking or recreational water safety can also lessen the quality of life for local communities.

BOX 2.8. CYANOHAB IN LAKE PONTCHARTRAIN TIED TO MISSISSIPPI RIVER DIVERSIONS INTO THE LAKE

In 1997, a massive bloom of *Anabaena cf. circinalis*, other *Anabaena* spp. and *Microcystis* sp. occurred in Lake Pontchartrain, a large lake (24 mi across) to the north of New Orleans, Louisiana. Microcystin concentrations in excess of WHO recommended levels resulted in an advisory against recreational use issued by the Louisiana Department of Health and Hospitals. The lake is normally slightly salty with low nutrients and chlorophyll. The bloom followed diversion of high nutrient Mississippi River water into the lake (to protect New Orleans from flooding) and was directly related to the increased nutrient loading and decreased salinity caused by the river water diversion (Dortch et al. 2001, Turner et al. 2004).

Aerial photograph of CyanoHAB on Lake Pontchartrain. Line across top is Causeway Bridge. *Photo: R.E. Turner, Louisiana State University*

Largemouth Bass impacted by *P. parvum* bloom in Possum Kingdom Reservoir (Texas). *Photo: Joan Glass, TPWD*

2.3. OTHER FRESHWATER HABS

Other, non-cyanobacterial, freshwater HABs cause harm by generating excessive biomass or through production of compounds that are toxic to fish or other aquatic organisms. Unlike CyanoHABs, they generally have had no known direct impacts on human health in the United States. These include some species of haptophytes, dinoflagellates, green algae, raphidophytes, euglenophytes, diatoms, and cryptophytes (Table 1). Although this report addresses freshwater HABs, some of these species bloom in inland waters of higher salinity, such as the Salton Sea.

Of the non-cyanobacterial inland HAB taxa, *Prymnesium parvum* (also referred to as "golden algae"), is likely the most problematic in U.S. waters. As reviewed by Watson (2001), *P. parvum* has caused large fish kills worldwide since as early as the 1930's, and was first suspected of fish kills in Texas in 1982 and confirmed in 1985. *P. parvum* generally blooms in brackish water and has killed millions of fish in Texas (Box 2.11). *P. parvum* has also been confirmed in New Mexico, Colorado, Wyoming, North Carolina, South Carolina, Georgia, Arkansas, Alabama, and Oklahoma and has been suspected in Florida and Nebraska. In New Mexico, *P. parvum* has entered into stream environments where it threatens the survival of the Pecos bluntnose shiner, which is a threatened fish species.

Nuisance species of the green algae, Chlorophyta, include both macro- and microalgae. For example, the macroalgal chlorophyte, *Cladophora*, has had a recent resurgence in the Great Lakes. *Cladophora* forms foul-smelling nuisance blooms that are deposited on beaches, can clog water intakes, and potentially harbor pathogens, such as *E. coli*. Economic impacts also result due to reduced beach use. Studies suggest that the resurgence of macroalgae in the Great Lakes may be due, in part, to increased water clarity caused by the introduction of the zebra mussel (Lowe and Pillsbury 1995). Other potential causes include possible increased nutrient inputs, increased water temperatures, and changing lake levels, although these factors need to be explored further (Bootsma et al. 2004).

Euglenophytes can be found in fresh, estuarine, and marine waters and are most common in slow moving or still waters with high nutrient concentrations. Blooms are most likely to occur during summer in freshwater ponds and ditches that receive nutrient-rich waste or runoff, and many species of euglenophytes are considered indicators of organic water pollution (Wehr and Sheath 2003). Euglenophyte blooms may color the water green, reddish brown, or red and are generally not toxic, but some freshwater euglenophytes have recently been implicated in fish kills in North Carolina, presumably due to the production of an ichthyotoxic compound (Zimba et al. 2004).

BOX 2.9. CYANOHABS IN THE POTOMAC RIVER AND OTHER CHESAPEAKE TRIBUTARIES

In the 1960's, the Potomac River below Washington D.C. had large outbreaks of CyanoHABs. Although improved wastewater treatment has reduced bloom occurrence since then, Maryland Department of Natural Resources (MD DNR) has documented cyanobacterial blooms almost annually in Chesapeake tributaries since 1985. An extensive *Microcystis* bloom on the Sassafras River at Betterton Beach in 2000 resulted in Maryland's first ever beach closure due to the presence of cyanotoxins (Tango et al. 2008). Beach closures occurred again in 2003 and health advisories were issued for various affected tributaries in 2004, 2005 and 2006.

In 2004, a CyanoHAB occurred in the Potomac River and several upper Chesapeake Bay tributaries in late spring through summer, with microcystin levels high enough to pose human health risks in some locations. In June and early July, the Virginia Department of Environmental Quality closed beaches due to high abundance of cyanobacteria, and MD DNR issued a press release cautioning citizens about recreating in affected waters. This massive CyanoHAB, which at one point covered a distance of approximately 30 miles, persisted until late summer. As the bloom declined, associated white foam covered portions of the Potomac River for nearly 2 miles.

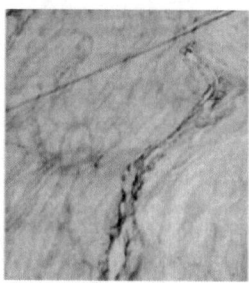

Foam accumulation in the Potomac River from a Microcystis bloom in 2004.
Photo: R.Lacouture, Morgan State University

Another bloom occurred but was smaller, shorter, and of lower cell concentration during summer 2005. Most recently, in August of 2006, a bloom of comparable extent to 2004 (26 miles) but shorter duration and lower magnitude maximum cell concentration 1% of 2004 maximum) occurred in the same area of the Potomac River. MD DNR responded quickly, acquiring microcystin toxin data in time to support State and county Health Departments in issuing a recreational advisory before the Labor Day weekend to protect human health.

BOX 2.10. CYANOHABS IN FLORIDA'S LAKES, RIVERS, AND UPPER REACHES OF ESTUARIES

Many of Florida's lakes, rivers, and upper reaches of estuaries are impacted by toxic CyanoHABs. Florida's largest lake, Lake Okeechobee, has experienced CyanoHAB outbreaks since the 1970's, with some of the largest covering over 100 square miles of the lake (reviewed in Phlips et al. 2002). In 1999, a statewide survey for cyanobacteria and cyanotoxins found over half the samples collected (representing 75 waterbodies) contained cyanotoxins, with microcystins being the most commonly occurring. In a similar survey in 2000, microcystins and cylindro spermopsin in some finished drinking water samples exceeded the current existing (microcystin-LR) and proposed (cylindrospermopsin) WHO drinking water guidelines (Burns et al. 2002).

In the summer of 2005, an unusually severe bloom occurred in the Lower St. Johns River. Large rafts of toxic algae were transported through the city of Jacksonville, and the Florida Department of Health released a public health advisory to deter recreational use of the river during the bloom event. Toxic CyanoHABs also occurred that summer in Lake Okeechobee, Caloosahatchee River, St. Lucie River, and the C-51 Canal (in West Palm Beach, Florida) (Burns 2008). No statewide monitoring program exists for CyanoHABs, but Florida Department of Health Aquatic Toxins Program responds to and issues health advisories for CyanoHABs that pose potential threats to public health. Some water districts have initiated monitoring efforts as well (see Appendix III).

BOX 2.11. IMPACTS OF *PRYMNESIUM PARVUM*, "GOLDEN ALGAE", ON FISHING IN TEXAS

Prymnesium parvum has been documented in more than 25 lakes and rivers in five of Texas' major river basins, with most toxic blooms occurring in winter months and in brackish inland waters. *P. parvum* blooms are estimated to have killed more than 17.5 million fish worth nearly $6.5 million in Texas lakes, rivers, and fish hatcheries between 1985 and 2003 (Glass 2003). A large fish kill at Possum Kingdom Lake, Texas, in 2001 was estimated to cause losses of $2.8 million for local economies due primarily to lost revenue from fishing-related tourism (Oh and Ditton 2005). *P. parvum* has caused fish kills in Texas each year since 2001. Fish kills have included game fish, such as largemouth bass, smallmouth bass, striped bass, catfish, crappie, and rainbow trout, and threatened species, such as blue suckers and Rio Grande darters. Blooms have likely further altered ecosystems by causing shifts in aquatic invertebrate and microorganism communities.

Fish kill due to P. parvum bloom on Lake Granbury in Texas. *Photo: Joan Glass, TPWD*

The marine raphidophyte, *Chattonella*, which has caused fish kills in Japan and Australia, has also been linked as a potential cause of fish kills in the Salton Sea, which is a saline but inland waterbody in California (SSERG

2001). The freshwater raphidophyte, *Gonyostomum semen*, and the freshwater dinoflagellate, *Peridinium polonicum*, form nuisance blooms in other countries, with the latter causing fish kills. These species have the potential to cause problems in the United States. *Peridinium polonicum* blooms have caused water discoloration in Oklahoma (Nolen et al. 1989), for example, but have not been associated with fish kills in the United States. *Didymosphenia geminata*, a species of freshwater diatom, is an emerging problem as it produces large amounts of extracellular stalk material that can dominate stream surfaces, altering hydrologic and biological conditions (Spaulding and Elwell 2007). Cryptophytes often form blooms that discolor the water but are not toxic and generally not harmful.

In: Marine and Freshwater Harmful Algal Blooms   ISBN: 978-1-60741-838-2
Editor: Peter E. Williams   © 2010 Nova Science Publishers, Inc.

Chapter 3

RESEARCH ON FRESHWATER HABS IN THE UNITED STATES

Joint Subcommittee on Ocean Science and Technology

U.S. Federal agencies are endeavoring to understand the fundamental science of freshwater HABs, to develop prevention and control strategies, and to mitigate effects on humans and wildlife. The study of freshwater harmful algae has been done at the individual project level as a part of larger research programs in a number of agencies. Coordination among Federal agencies and among Federal, state, and local agencies is improving but still needs strengthening. State and local agencies have a major role in monitoring blooms, researching treatment options, and alerting the public to health threats posed by HABs (see Appendix III for state activities). Current, new, and proposed research and activities to advance prediction and response for freshwater HABs are discussed in this chapter, with contributing agencies highlighted in blue, and detailed by agency in Appendix I. Information on the state of science for this chapter was gathered via: 1) a request for information from entities (see Appendices I-II) involved in various aspects of HAB research and response, including Federal agencies and the nonprofit organization, the American Water Works Association Research Foundation (AwwaRF), 2) a careful review of ISOC-HAB documents, 3) a general literature survey, and 4) public information on the web. In addition, this chapter describes how the current interagency extramural research funding program, ECOHAB, will satisfy the HABHRCA 2004 mandate to include freshwater HAB research.

U.S. freshwater HAB research and response builds on research and planning which has been ongoing internationally for the past two decades (see Appendix IV for international activities). For example, Australia has a nationally coordinated program targeting algal blooms in inland waters that focuses on toxic CyanoHABs. Germany, Switzerland, Denmark, Spain, Finland, Portugal, Poland, the Netherlands, and the United Kingdom are involved in a program, called TOXIC, to address CyanoHABs in European countries. CyanoHABs in drinking water sources are also a priority for the Global Water Research Coalition (GWRC), and there are a number of global coordination efforts (e.g. GWRC 2004, Codd et al. 2005). AwwaRF, CDC, and EPA have been important U.S. entities contributing to these international coordinated efforts.

3.1. CURRENT FEDERAL RESEARCH

3.1.1. Occurrence of Freshwater HABs

Before effective management can be implemented, it is important to understand the spatial and temporal distribution of HABs and HAB toxins. In 1996, AwwaRF began the process of assessing cyanotoxin (specifically microcystins) occurrence in drinking water sources in the United States (Carmichael 2001), but as the first nationwide survey, the project was limited in scope (see Section 2.2.1 for discussion of survey results). Several Federal agencies, including NOAA, CDC, EPA, USGS, and the U.S. Fish and Wildlife Service (USFWS), as well as many state agencies have recently been involved in monitoring waterbodies for both cells and, to a limited extent, toxins associated with CyanoHABs and other freshwater HABs. However, the information on occurrence of freshwater HABs and their toxins is fragmented in diverse databases and not readily accessible.

Regional monitoring of freshwater HABs has occurred in some parts of the country. Projects in multiple NOAA laboratories and programs—the Great Lakes Environmental Research Laboratory (GLERL), Center for Coastal Monitoring and Assessment (CCMA), Center for Coastal Fisheries and Habitat Research (CCFHR), Monitoring and Event Response for HABs Program (MERHAB), and the Oceans and Human Health Initiative (OHHI)—have worked to describe the occurrence and distribution of CyanoHABs and toxins in the Great Lakes and upper reaches of estuaries (e.g., Box 3.1). Investigators from GLERL and CCFHR and others funded by MERHAB have studied the distribution of CyanoHABs, microcystin toxins, and toxigenic strains in the Great Lakes, determining that some waterbodies experience microcystin levels exceeding WHO safety guidelines for finished drinking water (Box 2.5). Satellite imagery products provided by CCMA have helped guide sampling efforts by identifying potential blooms. The OHHI extramural program is also funding research to identify emerging freshwater HAB species and toxins that pose threats to human health in Lake Erie.

NOAA's MERHAB Lower Great Lakes Project is a multi-lake, comparative regional study focused on application of new tools to detect, track, predict, and respond to blooms of cyanobacteria and their associated toxins. The project is comprised of three field components in Lakes Erie, Ontario, and Champlain to investigate the spatial and temporal distribution of cyanotoxins and to determine the composition of the algal community across the region. Using quantitative real-time PCR, researchers highlighted a previously undocumented diversity of microcystin-producing cyanobacteria in the western basin of Lake Erie.

Box 3.1. NOAA Maps Toxin Concentrations in Great Lakes

NOAA CCFHR researchers, in collaboration with GLERL, are looking at occurrence of microcystin toxins in Saginaw Bay (Lake Huron) and in Lake Erie and are using this information to inform public health officials and environmental managers. They developed maps showing both microcystins inside the cyanobacterial cells (intracellular) and microcystins dissolved in the water (extracellular) for Saginaw Bay and western Lake Erie. They found that, in general, the most toxic cells in Saginaw Bay were located closer to shore. In western Lake Erie, the most toxic cells were generally found on the western and southern shores.

NOAA MERHAB also supports projects monitoring HABs and water quality in the Chesapeake Bay and St. John's River Estuary, Florida, which have developed real-time monitoring capabilities and documented the occurrence of CyanoHABs in the upper reaches of these estuaries. An EPA Region 4 project in the upper Barataria Basin in Louisiana is testing for toxins in water and in crustaceans and shellfish eaten by humans. Other states, including Nebraska, Florida, Indiana, Maryland, Oregon, Texas, Vermont, and Washington, have state programs that monitor targeted lakes and other recreational waterbodies for cyanobacteria and other toxic HABs. Vermont has a volunteer monitoring program.

The spread of the golden alga, *Prymnesium parvum*, from Texas rivers into New Mexico is being watched closely by USFWS in New Mexico. The State of New Mexico Department of Game and Fish has been conducting water testing to document the occurrence of *P. parvum* associated with fish kills. Texas Parks and Wildlife Department (TPWD) has documented blooms of *P. parvum* in at least 5 Texas river basins. USFWS has also contributed funds to the TPWD to expand its understanding of golden alga.

USGS is monitoring water quality and occurrence of CyanoHAB events on a broader scale. The Columbia Environmental Research Center of the USGS is involved in monitoring inland waterbodies, including documentation of microcystins and other toxins in reservoirs and wetlands. The USGS National Wildlife Center analyzes samples associated with wildlife poisonings and maintains a database documenting events, such as mass bird mortalities, to record both CyanoHAB occurrence and ecological effects. Through the National Water Quality Assessment Program (NAWQA), USGS is also conducting ecological studies and long-term, widespread sampling of algae to relate blooms to water quality conditions. The USGS implemented NAWQA in 1991 to develop long-term consistent and comparable information on streams, rivers, ground water, and aquatic systems in support of national, regional, state, and local needs for information and policy determinations related to water quality management.

Infrastructure, such as culture facilities that maintain collections important for developing new detection methods and for training in taxonomic identification, enable efforts to document occurrences of freshwater HABs. The National Science Foundation (NSF) provides ongoing support to the University of Texas (UTEX) culture collection, which contains *P. parvum* and some toxin-producing cyanobacteria cultures. NSF is also supporting the development of Reactive Observing Systems, which involve *in situ* real-time sampling by stationary (12 buoys) and mobile units (robotic boats). These systems can monitor concentrations of algae and other microorganisms in waterbodies and are able to reconfigure their sampling routine in response to observations collected. These systems are being tested currently in Lake Fulmar, California.

3.1.2. Understanding Bloom Dynamics

An understanding of bloom dynamics (e.g., how environmental conditions promote HABs) is critical for explaining their causes, developing prevention strategies (e.g. Box 2.2), and predicting their occurrence. Past research has shown that nutrient and hydrologic conditions have strong impacts on the development and maintenance of CyanoHABs, but interactive roles of these and other factors (e.g., organic matter availability, light attenuation, temperatures, grazing, bacterial and viral infection, life cycles) on bloom dynamics and toxin production are difficult to characterize (Perovich et al. 2008). Improved understanding of these complex interactions will lead to predictive models and allow development of successful strategies for HAB prevention and control. Finally, research on causes of other, non-cyanobacterial freshwater HABs in the United States has been considered a lower priority to date.

NOAA GLERL is investigating the influence of environmental factors on *Microcystis* blooms, and NOAA and EPA are supporting ECOHAB research examining the interactions between grazers, nutrients, and HAB events. The grazers studied are the invasive zebra mussel whose presence in the Great Lakes has been associated with proliferation of the cyanobacteria, *Microcystis aeruginosa*. EPA's Office of Research and Development is also conducting research on the role of nutrient loading in algal blooms. NSF supports research on important aspects of cyanobacterial biology and ecology, including elucidating metabolic and regulatory networks, analysis of nitrogen fixation in some species, vertical distribution in lakes using mathematical models, and the role of dissolved organic material in regulating primary production in prairie lakes.

3.1.3. Prevention

Prevention strategies are important for management of freshwater HABs, and successful strategies are built on an understanding of the causative factors (Section 3.1.2). The USGS monitors streamflow, nutrient, and other water quality parameters that may have important impacts on freshwater HAB dynamics. Water quality monitoring of inland watersheds, used together with model analyses, is important to define source locations of nutrient loadings and human and natural factors that influence these loads. These data are useful for managing water quality impacts, such as eutrophication, that can lead to HABs.

> **BOX 3.2. USDA PHOSPHORUS INDEX CONTRIBUTES TO REDUCED PHOSPHORUS LOADINGS**
>
> USDA Agricultural Research Service developed the Phosphorus Index, which is a phosphorus assessment tool that has been adopted by USDA's Natural Resources Conservation Service, EPA, and state agencies as the basis for their comprehensive phosphorus management plans. The adoption of this technology has been estimated to reduce phosphorus loadings in water by 56 million pounds, and sediment by 2.1 billion pounds, annually. Economic benefits to society have been estimated at greater than $600 million dollars per year.

Since excessive nitrogen and phosphorus can cause HABs in downstream aquatic ecosystems, USDA is researching how to reduce nitrogen and phosphorus loading to surface water from agricultural lands through best management practices. Most of this extramural research is focused on marine and estuarine HABs, but one of the projects focuses on how water quality is related to occurrence and abundance of freshwater HABs in the St. Johns River, Florida. USDA developed the Phosphorus index tool (Box 3.2), which has reduced phosphorus loadings in water. USDA is also supporting laboratory investigations into the use of wetlands to treat swine wastewater, which has high nitrogen and phosphorus content and the potential to cause HABs in freshwater receiving systems. Finally, USDA's best management practices for controlling off-flavor problems in catfish ponds include changes in rotation of fish crops and complete flushing of ponds every few years in an effort to prevent blooms from occurring.

3.1.4. Mitigation and Control

Drinking water, recreational waters, and aquaculture facilities have suffered negative impacts from freshwater HABs, most notably from CyanoHABs. These negative impacts can be lessened through *mitigation* efforts, which include monitoring (see Section 3.1.1), toxin removal, predictive modeling to provide early warning, and event response. Mitigation can be enhanced by better understanding impacts (see Section 3.1.6), leading to improved outreach and education. *Control*, or the direct reduction or removal of an existing bloom, can prevent negative impacts and potentially be carried out via biological, chemical, or physical means. Some efforts to control HABs and improve toxin removal in drinking water reservoirs and water treatment facilities are underway both by Federal agencies and by state and local authorities. Predictive capabilities for CyanoHABs in reservoirs, lakes, and rivers are also in development by state and Federal agencies. Prediction of onset and control of CyanoHABs in aquaculture facilities is being explored, as well, to help mitigate economic losses experienced as a result of off-flavor in catfish caused by cyanobacteria, some of which are HAB species. Outreach and education to inform the public of dangers of HABs in recreational waters and to alert the public when events occur is being conducted primarily by state agencies. Specific efforts in mitigation and bloom control are discussed below.

BOX 3.3. SEA GRANT RESEARCHERS DEVELOP METHODS TO REMOVE TOXINS FROM DRINKING WATER

Water treatment facilities do not specifically treat drinking water for microcystins and many of the conventional removal processes are ineffective on these toxins. Sea Grant researchers in Lake Erie have developed an efficient method that removes up to 95% of harmful microcystins from drinking water using a combination of wood-based powdered activated carbon and ultrafiltration technologies. Many water treatment facilities already use powdered activated carbon to remove herbicides and taste-and-odor compounds, so the combination of this approach with ultrafiltration offers an efficient and cost-effective method for removal of microcystins.

3.1.4.a. Mitigation for Algal Toxins in Drinking Water

EPA promulgates Federal clean water and safe drinking water regulations, standards, criteria, and guidance and issues health advisories. The Agency provides support for municipal wastewater treatment plants and takes part in pollution prevention efforts aimed at protecting watersheds and sources of drinking water. The Agency oversees both regulatory and voluntary programs to fulfill its mission to protect the nation's waters. EPA has 10 regional offices that generally act as liaison between state and Federal authorities. If a problem occurs in waterways, the regional offices respond and relay information about the event to EPA's Office of Water. EPA has also partnered with the AwwaRF (http://www.awwarf.org/) to create guidance on methods for controlling and mitigating algal growth within water treatment plants and to develop criteria for laboratories providing algal toxin analyses to drinking water utilities. While drinking water (also referred to as tap water) is regulated by EPA, bottled water is regulated by FDA. As with drinking water, there are currently no standards or guidelines for cyanobacteria or cyanotoxins in bottled water. More information regarding FDA's regulation of bottled water can be found at http://www.cfsan.fda.gov/~dms/botwatr.html. Since no Federal regulations exist for setting standards for cyanotoxin levels in drinking water, states are not mandated to test for cyanotoxins, but some are doing so proactively. Current testing is mainly limited to microcystins, the only class of cyanotoxins for which an enzyme-linked immunosorbent assay (ELISA) method has been standardized.

Standard treatment procedures in drinking water treatment facilities can remove or inactivate some algal toxins, but additional treatment technologies improve toxin removal efficiency and are often necessary. Westrick (2008) suggested the importance of considering specifics at each treatment facility for determining the most effective treatment for drinking water. The efficiency of toxin removal depends on the toxin type and whether it is dissolved in the water (extracellular) or inside cyanobacterial cells (intracellular), with extracellular toxins being the most difficult to remove. More utilities are now using more than just the conventional treatment of flocculation, coagulation, sedimentation, and filtration alone, which generally does not effectively remove extracellular toxins unless the toxins are oxidized through disinfection. Moreover, data from Florida have demonstrated that cyanotoxin levels in finished drinking water can exceed those in raw water, presumably because the cells may rupture during filtration causing release of the intracellular toxins. The disinfectant chlorine is now being replaced by other oxidizing disinfectants, some of which (e.g. chloramine, chlorine dioxide, ultraviolet disinfection) have not shown promise for degrading cyanotoxins (Westrick 2008). In addition, the complex interactions between the specific oxidant, the toxins present, environmental factors (such as total organic compounds, temperature, pH and contact time) that determine the efficiency with which HAB toxins are oxidized are not well understood, and the toxicity of the degradation products are unknown.

A few technologies are currently in development by Federal agencies for improving removal of toxins from drinking water. Most studies into water treatment methods have focused on microcystins. Researchers from NOAA's National Sea Grant College Program have developed a new technique that has been shown to remove 95% of microcystins from drinking water (Box 3.3). EPA's NERL is exploring the use of an emerging "green" technology for the same purpose (Box 3.4). The cities of Cocoa and Melbourne, Florida and St. John's Water Management District in Florida are conducting a research study, in coordination with the company, CH2M HILL, and with funding from AwwaRF, to characterize various treatment technologies for removal of algal toxins from drinking water supplies. The treatments considered include ozone, advanced oxidation, powdered and granular activated carbon, biological treatment, and membrane filtration. Another study funded by AwwaRF with the City of Phoenix Water Services Department and Arizona State University, developed a monitoring and response program to rapidly assess emerging problems in distribution systems related to algal toxin and taste-and-odor issues. In addition, the Metropolitan Water District of Southern California is doing research for predicting and managing taste-and-odor problems caused by CyanoHABs in drinking water.

BOX 3.4. EPA IS TESTING NEW "GREEN" TECHNOLOGY FOR TOXIN REMOVAL

The National Exposure Research Laboratory (NERL) of EPA has completed important research on the application of "green" technology for the treatment of microcystins in drinking water. Pilot studies have shown that conventional treatment processes such as coagulation, flocculation, and sedimentation result in increased levels of soluble toxin. A promising chemical oxidation technology for the treatment of microcystin –LR is titanium dioxide photocatalysis. This technology efficiently performs water purification and disinfection. This study showed that immobilized titanium dioxide photocatalysis could effectively destroy microcystin-LR in water at concentrations up to 5000 $\mu g\ L^{-1}$.

Monitoring of source waters and drinking water intakes is also important for mitigation of algal toxins in drinking water. EPA's National Center for Environmental Research (NCER) is currently supporting research in the academic community to develop gene microarray assays for monitoring cyanobacteria and cyanotoxins in drinking water. This study will produce a microarray suitable for use as a tool to assess cyanobacteria and cyanobacterial toxin risks in drinking water reservoirs and lakes. See Section 3.1.1 for other monitoring activities. Bloom control methods (i.e., to eliminate or contain blooms) are discussed in Section 3.1.4.c.

> **BOX 3.5. NEW HARVESTING TECHNOLOGY IMPROVES SAFETY OF BLUE-GREEN ALGAL SUPPLEMENTS**
>
> Simplexity, formerly Cell Tech International, harvests and sells blue-green algal (*Aphanizomenon flos-aquae*) nutritional supplements from Upper Klamath Lake in Oregon. During routine quality control testing in the 1990's, it was determined that the co-occurring cyanobacterium, *Microcystis*, was present in the lake algae population at certain times of the year. *Microcystis* produces the toxin microcystin (Table 2). The company worked closely with engineers and scientists to design a monitoring and harvesting system that would ensure the safety of their product. Using an antibody–based testing kit and routine microscopic examination of plankton samples, managers are able to monitor the relative abundance of *A. flos-aquae* and *Microcystis* and to decide when to harvest. An innovative floating pontoon-based harvesting system was then designed and tested that effectively removes contaminating *Microcystis* while retaining the desirable *A. flos-aquae*. With on-board refrigeration and other quality control procedures in place, the company is now able to harvest high-quality product that fully complies with state and federal guidelines. This is a real-world example of how effective HAB mitigation strategies can maintain healthy industries in areas subject to recurrent blooms.
>
>
>
> Self propelled floating platform for harvesting *Aphanizomenon flos-aquae* on Upper Klamath Lake. Note shuttle craft (left) for quickly moving freshly harvested algae to the processing facility. *Source: Simplexity*

3.1.4.b. Mitigation for Algal Toxins in Algal Nutritional Supplements

Toxins in nutritional supplements containing cyanobacteria are also a potential problem (Lawrence et al. 2001). Because FDA is responsible for the safety of nutritional supplements, that agency has been analyzing for the toxicity of supplements contaminated from potentially co-occurring toxic cyanobacteria. Private industry has also been involved in the development of new monitoring and harvesting technologies to prevent microcystins in nutritional supplements (See Box 3.5).

3.1.4.c. Bloom Control

Types of potential control for freshwater HABs include artificial destratification, increased flushing rates, algicidal compounds, coagulation (e.g. using clay), ultrasound, and biological manipulation (Paerl 2008, Perovich et al. 2008). Destratification, which mixes the water column vertically, has had limited success and is feasible only in smaller waterbodies. Increased flushing is only possible where upstream water supplies are available. Algicides are a common method of control in ponds, reservoirs, and aquaculture facilities, but can have unintended impacts. First, algicides can cause the algal cells to lyse, releasing toxins into the water and potentially creating a high risk situation. Second, copper sulfate, the most commonly used algicidal compound in smaller waterbodies, like other copper-based chemicals, is toxic to all flora and fauna so its use is not advisable in many cases (Paerl 2008). Studies to characterize the levels and effects of algicides and HAB toxins in aquatic food webs, which include fish consumed by humans, have not been conducted.

Currently the only approved chemicals for use in catfish aquaculture ponds are copper-based algicides and the herbicide diuron, which are undesirable because of the broad spectrum of toxicity and their persistence in the environment. Some naturally derived compounds, such as a by-product of aged barley straw (Barrett et al. 1999), are potential alternatives that exhibit more selective toxicity. USDA is conducting research to develop new control methodologies, including an algicidal compound derived from ryegrass, to selectively eliminate cyanobacteria from phytoplankton communities inhabiting ponds used for catfish aquaculture. In addition, USDA has been experimenting with detecting toxin-producing algae in catfish aquaculture facilities with low-altitude remote sensing methods which may assist in targeting timely control efforts. The USACE is also evaluating possible new chemicals, other than copper-based algicides, to control blooms in drinking water reservoirs and is developing a framework to manage blooms when they occur in the 400 reservoirs for which USACE is responsible.

TPWD is working with researchers, other agency officials, and interested parties within and outside of Texas to better understand and potentially control harmful golden algae (*Prymnesium parvum*) in Texas aquaculture facilities. Clay flocculation (Sengco et al. 2005) and extracts of barley straw (Roelke et al. 2007), which have been successful in controlling some HABs, have not shown promise for controlling *P. parvum*, but TPWD fish hatcheries have been able to successfully treat their ponds with ammonium sulfate, copper sulfate, other copper-based algicides, ozone, and ultraviolet radiation. Researchers at the USDA Western Regional Research Center have also been examining the use of copper to control filamentous algae (chlorophyta, cyanobacteria) in rice fields.

3.1.4.d. HAB Prediction

Prediction of HAB events can range from simple models that predict the likelihood of bloom occurrence or transport based on a few easily measured variables to complex models that incorporate biological, chemical, physical, and hydrological factors. The Chesapeake Bay Program and MD DNR currently issue a Summer Ecological Forecast based on a simple model that uses river outflow to project the likelihood of *Microcystis* blooms in the Potomac River. A current NOAA MERHAB project is supporting research to advance these predictive capabilities. The USGS Kansas Water Science Center has developed monitoring tools to predict the onset of cyanobacteria-related taste-and-odor episodes in drinking water reservoirs based on continuously monitored light, temperature, conductivity, and turbidity (*http://ks.water.usgs. gov/Kansas/studies/qw/cyanobacteria/*), and ongoing studies are working to improve these predictions and develop capabilities to predict other variables of concern, such as cyanotoxins. USDA scientists have been researching the relationship between the occurrence of cyanobacteria and off-flavor in catfish in aquaculture ponds and developing molecular biological methods to allow forecasting of problematic periods and implementation of best management practices.

There are currently two efforts underway to develop a system for predicting bloom transport in the Great Lakes: one by NOAA's GLERL and CCMA which may develop into a forecasting system, such as one already operationalized for the west coast of Florida, and a second, funded by NOAA MERHAB's Lower Great Lakes project, which focuses on Lakes Erie, Ontario, and Champlain. GLERL is developing the use of remote detection of cyanobacteria through phycocyanin pigment analysis and has already developed a hydrodynamic model to predict bloom transport in the Great Lakes. The MERHAB transport model couples a particle tracking model and a hydrodynamic model (a version of the Princeton Ocean Model, http://www.aos.princeton.edu/WWWPUBLIC/htdocs.pom/) to predict bloom transport. As with weather prediction, the use of multiple models, with different assumptions, improves forecasting accuracy.

BOX 3.6. NOAA GLERL HAB WEBSITE PROMOTES ACTIVE RESEARCH COLLABORATION

The Harmful Algal Bloom Event Response website (http://www.glerl.noaa.gov/res/Centers/HABS/habs.html) has been a successful outcome of the NOAA's Center of Excellence for Great Lakes and Human Health and the research programs at GLERL. Researchers are combining ground-based measurements and satellite image data to characterize bloom dynamics and inform development of future bloom forecasting tools. As a forum to post data, the website fosters a productive collaboration for researchers studying CyanoHABs in the region and for the public.

BOX 3.7. SAMPLE GUIDELINES FROM THE WISCONSIN DIVISION OF PUBLIC HEALTH'S FACT SHEET ON CYANOBACTERIA AND THEIR TOXINS AND HEALTH IMPACTS (PEROVICH ET AL. 2008)

http://dhfs.wisconsin.gov/eh/Water/fs/Cyanobacteria.pdf
- Never drink untreated surface water, whether or not algal blooms are present. Boiling the water will not remove toxins. Owners should always provide alternative sources of drinking water for domestic animals and pets, regardless of the presence of algae blooms.
- If washing dishes in untreated surface water is unavoidable, rinsing with bottled water may reduce possible residues.
- People, pets and livestock should avoid contact with water where algae are visible (e.g., pea soup, floating mats, scum layers, etc.) or where

> the water is discolored. Do not swim, dive, or wade in this water. Do not use the water to fill a pool or for an outdoor shower.
> - Always rinse off yourself and your pet after swimming in any ponds, lakes or streams, regardless of the presence of visible algae blooms. Pay close attention to the bathing suit area and pet's fur.
> - Contact your local health department or department of natural resources office to report any large algae blooms on public or private lakes, streams or ponds.
> - Never allow children or pets to play in or drink scummy water. Do not allow pets to eat dried scum or algae on the shoreline.
> - Do not water-ski or jet-ski over algae mats.
> - Do not use algicides to kill the cyanobacteria. When the cyanobacteria cells die, the toxins within the cells are released.
> - Obey posted signs for beach closings. Wait at least one to two weeks after the disappearance of cyanobacteria before returning to the water for wading, bathing or other activities.

3.1.4.e. Outreach

Outreach enhances mitigation efforts by promoting community awareness of HAB issues, event occurrence, and potential health threats and by dispelling misperceptions of water safety. Some states have produced short informational documents that are available on the internet or disseminated via kiosks near impacted recreational areas (e.g., Box 3.7). Seventeen states have websites informing the public about the negative impacts of freshwater HABs and how to respond when they occur. In each state, one or more state agencies are involved in alerting the public to HABs in lakes or ponds where recreational exposure may occur. A few states have active hotlines for reporting problems, including Florida, Maryland (web form), New Hampshire, Vermont, and Nebraska. Oregon, Maryland, Nebraska, and Vermont provide the public with very specific advisories and maps showing the location of blooms. Some states, including Florida, Texas, and Nevada, have task forces addressing the issue of HABs. See Appendix III for details on state activities.

Outreach activities by Federal agencies have included the education of state and local officials and dissemination of data and information via websites. NOAA's GLERL has created a website and list server to distribute data about CyanoHABs in Lake Erie to other researchers and the public (Box 3.6). NOAA's MERHAB-funded workshops in New York and Vermont helped regional water treatment facility operators and public health officials learn how to recognize cyanobacterial problems in drinking water and how to minimize the public health effects.

3.1.5. Toxin Research and Methods for Detection and Analysis

Fast, accurate, and reliable methods for toxin detection in the field are important for protecting human health and furthering research (HARRNESS 2005, Hudnell 2008). The development of methods for toxin identification and quantification is dependent on the availability of high quality toxin standards and standardized protocols for toxin extraction, concentration, and analysis in various matrices, such as water and biological tissues (HARRNESS 2005, de la Cruz et al. 2008).

In the United States, the U.S. Army Medical Research Institute of Infectious Diseases (USAMRIID), NOAA, USGS, FDA, and EPA are involved in developing new analytical methods and facilities for toxin analysis. USAMRIID is developing targeted diagnostic capabilities to detect exposure to microcystins and saxitoxin in humans. FDA not only serves as a source of reference standard for saxitoxin, but is also involved in methods development for saxitoxin detection, and to a lesser extent, microcystins. NOAA's Oceans and Human Health Center at the GLERL is developing new analytical tools for studying emerging freshwater HAB cells and toxins, including establishing molecular and biochemical diagnostics for the presence of cells and their associated toxins, generating a series of toxin and cell standards to be used by GLERL, and evaluating the use of high-throughput assays to detect cyanobacterial toxins. Several regional USGS centers are developing sample collection and analytical techniques for measuring cyanotoxins. The establishment of the toxin analysis facility at the State University of New York College of Environmental Science and Forestry (SUNY-ESF), partly funded by NOAA's MERHAB Program, has enhanced overall abilities to identify and assess toxin distributions in the Great Lakes and elsewhere. This facility can analyze for all known classes of cyanobacterial toxins.

EPA (National Exposure Research Laboratory, or NERL; NCER; and Office of Water) has supported intramural and extramural research on new technologies for toxin detection and has contributed to the development of state-of-the-art, high-efficiency techniques for separation, detection, identification, and quantitative measurement of six, high-priority cyano bacterial toxins including anatoxin-a, cylindospermopsin, and four microcystins (microcystin-RR, -LR, -YR, and -LA). This research includes development of gene microarray assays for monitoring of cyanotoxins, especially in reservoirs and lakes that supply drinking water. EPA has also funded Phase I of a Small Business Innovation Research (SBIR) project to develop a faster, simpler fiber optic probe for detecting microcystin-LR in the field.

Progress has been made by scientists worldwide toward identifying genes involved in cyanotoxin synthesis, and this knowledge is important for developing molecular-based screening assays to detect active toxin production as well as to understand environmental factors that promote toxin production. NOAA's CCFHR is conducting genetic analysis of *Microcystis* strains from the Great Lakes to determine occurrence of toxigenic strains, which will help determine factors controlling bloom toxicity.

> ## BOX 3.8. CDC LAUNCHES HAB ILLNESS SURVEILLANCE SYSTEM
>
> The five coastal states that have been collaborating with the CDC's National Center on Environmental Health (NCEH) on HAB issues are currently participating in beta-testing of a new surveillance system. The original surveillance system for estuarine-associated illnesses was developed in response to the perceived public health threat from *Pfiesteria piscicida*. The new Harmful Algal Bloom-related Illness Surveillance System (HABISS) is designed to capture human and animal health data as well as environmental data for HABs in fresh, estuarine, and marine waters. HABISS is located on NCEH's Rapid Data Collection platform. The system is web-based and currently has seven modules for data entry, including physical information about the bloom event and medical information about algal toxin-related human and animal illnesses. NCEH hosted a HABISS workshop to expand the capability to an additional five states with HAB issues. It is anticipated that, after a few years of data collection, HABISS will allow state and local health agencies to forecast where blooms are likely to occur and implement the relevant exposure-mitigation and illness-prevention strategies to protect public health.

NOAA's MERHAB Program has supported similar work in Lake Erie and Ontario (e.g., Rinta-Kanto and Wilhelm 2006). In order to further the knowledge base of cyanotoxin synthesis, NOAA, through ECOHAB, supports a research project focused on identifying saxitoxin genes in the cyanobacterium *Anabaena circinalis*.

3.1.6. Human Health and Ecological Effects

3.1.6.a. Human Health Effects
An overview of human health effects of freshwater HABs is provided in Chapter 2 and Table 2. In the United States, research and response to mitigate the human health effects of freshwater HABs has included general tracking of HAB-related illnesses in five states; response to specific, possible poisonings; epidemiological research related to recreational exposure to HABs; toxicology research to further understand the effects of cyanotoxins; and exploration of the relationship between algal mats and harmful bacteria.

Observational epidemiological studies allow investigation of human health effects due to "environmentally relevant exposures to naturally occurring mixtures of toxins," but such studies can be difficult to implement because knowledge of cyanotoxin occurrence is essential to document exposures (Hilborn et al. 2008). CDC is currently conducting epidemiology studies to assess human exposures to cyanobacteria. The CDC's National Center on Environmental Health has established cooperative agreements with five coastal states (and will add more as funding allows) to develop public health response plans for toxic freshwater blooms. Additionally, CDC is conducting studies to assess the acute health effects from recreational exposure to cyanotoxins in freshwater. They have developed the Harmful Algal Bloom-related Illness Surveillance System (HABISS), which is web-based with seven modules for data entry, including biological and physical information about the bloom event and medical information about algal toxin-related human and animal illnesses (Box 3.8). Centralized HAB data collection will enhance mitigation strategies and improve the ability to detect future HAB-related illnesses. CDC in cooperation with Mote Marine Lab and NOAA's GLERL has performed epidemiological studies on the human health effects from recreational exposure to toxic cyanobacteria at Bear Lake, Michigan.

The National Institute of Environmental Health Sciences (NIEHS) has a new project involving developmental toxicity assays that utilize zebrafish

(*Danio rerio*) embryos; this project will also assess the toxicity of cyanobacterial lipopolysaccharides using the zebrafish embryo model. NOAA's Sea Grant and USGS both have projects researching the link between fecal bacteria, such as *E. coli,* and nuisance algal mats, such as mats formed by the green algae *Cladophora*, which, when present, can trigger beach closings due to the negative health effects.

3.1.6.b. Ecological Effects

Research by Federal agencies into the ecological impact from freshwater HABs has primarily focused on studying effects of CyanoHABs on fish and birds. Although ecological effects of CyanoHABs are well documented (see Chapter 2 for description), the relative role of cyanotoxins compared to other bloom-induced stressors is not well-known (Fournie et al. 2008). NOAA, through ECOHAB, is funding a new project which examines how cyanobacterial toxins bioaccumulate in fish and affect survival, reproduction, and other indicators of species health. The USGS National Wildlife Center also analyzes samples stemming from wildlife poisonings and maintains a database documenting events such as mass bird mortalities. USFWS, as the principal Federal agency charged with protecting and enhancing the populations and habitat of more than 800 bird species, has documented toxic CyanoHAB events involving birds including the collection of carcasses and water samples for toxin analysis. The USGS Western Fisheries Research Center's Klamath Field Station, in partnership with the U.S. Bureau of Reclamation (BOR), is investigating the impact of algal blooms on aquatic species in Upper Klamath Lake, California. This research may assist the BOR and USFWS to protect endangered fish species in the region.

3.1.6.c. Human and Ecological Risk Assessment

Risk assessments are used to evaluate impacts of toxins on humans and ecosystems which are important for management decisions. Risk assessments integrate hazard identification, hazard characterization (qualitative and quantitative evaluation of effects), and exposure assessment (qualitative and quantitative evaluation of exposure risk) to estimate the potential risk (reviewed by Falconer 2005) and develop health guidelines. Guideline values for drinking water are determined based on cancer and noncancer toxicity, so sufficient data to show dose-response relationships are important (Falconer 2005). For example, the International Agency for Research on Cancer recently changed the classification of microcystin-LR to "possibly carcinogenic to humans" based on epidemiological data from China (IARC 2006).

Risk and impact assessment for exposure of some cyanotoxins is conducted by EPA's National Center for Environmental Assessment (NCEA). The USACE, in collaboration with EPA's NCEA and National Health and Environmental Effects Research Laboratory (NHEERL), is also assessing the toxicity and health risks of algal toxins. The EPA NCEA has recently prepared draft *Toxicological Reviews of Cyanobacterial Toxins: Anatoxin-a, Cylindrospermopsin and Microcystins (LR, RR, YR and LA)* as a series of dose-response assessments to support the health assessment of unregulated contaminants on the CCL (see Appendix I for web links to documents). The purpose of these documents is to compile and evaluate the available data regarding toxicity of these toxins to aid the EPA's Office of Water in regulatory decision making.

3.2. NEW OR PROPOSED RESEARCH ON FRESHWATER HABS

Besides the ongoing Federal efforts described in the sections above, several Federal agencies are planning projects to advance freshwater HAB research. Some of these activities are listed below.

EPA

- Development of new detection methods for cyanobacterial cells and toxins, such as microcystins, cylindrospermopsin, and anatoxin-a in drinking water. Reliable detection methods would enable EPA to add freshwater harmful algae and their toxins to the agents monitored in drinking water through the Unregulated Contaminant Monitoring (UCM) Program. EPA uses this program to collect occurrence data for contaminants suspected to be present in drinking water, but which do not have health-based standards set under the SDWA. The list is reviewed every five years. Research on the health effects of exposure via drinking water to cyanobacterial toxins, such as microcystins, cylindrospermopsin, and anatoxin-a.

CDC

- Assessment of human exposures to HAB toxins during recreational activities such as swimming in lakes with HABs.
- Participation in a multi-agency, international (with Australia) group to study the public health effects of exposure to cyanobacterial toxins.
- Use of data from the Great Lakes Observing System component of ocean observing systems to assess how waterborne risks affect lakeshore communities.

NOAA

- Continuation of *Microcystis* toxin production monitoring in the Great Lakes, which can serve as a data source for the exposure assessment component of epidemiological studies.
- Development of enhanced molecular tools for assessing the ability of *Microcystis* blooms to produce toxins.
- Development of accurate satellite monitoring capabilities to monitor and track blooms.
- Proposing (by GLERL) to fund and conduct additional research on *Microcystis,* including:

Extramural projects

- Research the genomics of microcystin synthetase genes
- Study the sublethal impacts of microcystin accumulation in fish

Intramural projects

- Determine which environmental factors promote cyanobacterial HABs
- Examine how zebra mussels promote blooms
- Develop forecasting technologies and detection tools

3.3. CURRENT AND FUTURE ROLE OF ECOHAB IN FRESHWATER HAB RESEARCH

The 2004 reauthorization of HABHRCA calls for this scientific assessment to "establish priorities and guidelines for a competitive, peer-reviewed, merit-based interagency research program as part of the Ecology and Oceanography of Harmful Algal Blooms (ECOHAB) [program], to better understand the causes, characteristics, and impacts of harmful algal blooms in freshwater locations..." (Public Law 108-456). The ECOHAB Program, as it is currently constituted, meets the requirements of the legislation and will continue to do so, depending on the availability of funding and the priorities of the participating agencies.

ECOHAB is a multi-agency partnership between NOAA's Center for Sponsored Coastal Ocean Research (CSCOR) and National Sea Grant College Program, the EPA, NSF, National Aeronautics and Space Administration, and the Office of Naval Research. Extramural research funded by ECOHAB is guided by priorities described in several reports: *The Ecology and Oceanography of Harmful Algal Blooms: A National Research Agenda* (ECOHAB 1995), the *Harmful Algal Research and Response: A National Environmental Science Strategy 2005-2015* (HARRNESS 2005), the *National Assessment of Efforts to Predict and Respond to Harmful Algal Blooms in U.S. Waters* (Jewett et al. 2007), and this *Scientific Assessment of Freshwater Harmful Algal Blooms*. Through a combination of long-term regional ecosystem studies and short-term targeted studies, ECOHAB seeks to produce new, state-of-the-art detection methods for HABs and their toxins, to understand the causes and dynamics of HABs, to develop forecasts of HAB growth, transport, and toxicity, and to predict and ameliorate their impacts on higher trophic levels and humans. Research results are used to guide management of coastal resources to reduce HAB development, impacts, and future threats. Projects selected for support must successfully compete in a rigorous external, peer-review process that ensures a high level of scientific merit. Projects involve a mix of investigators from academic, state, Federal, nonprofit, and private institutions. The multi-agency nature of ECOHAB promotes information sharing and coordination among the agencies. During the last three years, NOAA and EPA have funded projects in the Great Lakes and upper reaches of estuaries.

The NOAA MERHAB Program, which is similar to ECOHAB except it is not interagency, focuses on developing partnerships and improving monitoring to minimize and prevent impacts of HABs. It has also funded projects in the Great Lakes and upper reaches of estuaries. In the past, the combined funding spent on freshwater research provided by NOAA and EPA via the interagency ECOHAB Program and the NOAA MERHAB Program has exceeded $1 million per year.

Given the geographic mandates of some of the agencies involved, ECOHAB and MERHAB focus on freshwater HABs in the Great Lakes and upper reaches of estuaries rather than freshwater locations further inland. However, it continues to be important to advance research on inland HABs because the causes, characteristics, and impacts of inland HABs may differ from HABs in the Great Lakes and the upper reaches of estuaries.

In: Marine and Freshwater Harmful Algal Blooms   ISBN: 978-1-60741-838-2
Editor: Peter E. Williams   © 2010 Nova Science Publishers, Inc.

Chapter 4

PLAN FOR IMPROVING FEDERAL RESPONSE TO FRESHWATER HABS: RESEARCH AND INFRASTRUCTURE PRIORITIES

Joint Subcommittee on Ocean Science and Technology

This chapter identifies goals and priorities for advancing freshwater HAB research. These goals and priorities, along with strategies to improve coordination (discussed in Chapter 5), constitute a plan to provide guidance for Federal and other research and grant programs. The plan is meant to offer guidance for coordinating current activities and developing new research directions that may lead to improved management strategies for freshwater HABs. It should be noted that this report does not attempt to make any statements about the relative importance of freshwater HAB research compared to other research areas or other priorities for Federal or state investment.

4.1. PROCESS FOR DEVELOPING GOALS AND PRIORITIES

During the interagency ISOC-HAB (Hudnell 2008, see Chapter 1 for description), invitees participated in seven workgroups to identify research options in their areas of scientific expertise for advancing future research and response programs to address the growing CyanoHAB problem. Members of the IWG-4H (Box 1.2) considered research directions identified by each of the ISOC-HAB workgroup reports as well as priorities for non-cyanobacterial freshwater HAB research when developing priorities and goals for guiding a national research strategy for freshwater HABs.

Seven priorities for advancing freshwater HAB research and response are presented in Table 3 and discussed in Section 4.2. For each priority, a number of more specific goals are provided. Some specific goals were identified as "immediate goals" to indicate a critical step for the corresponding research priority. "Immediate goals" share the following attributes:

- identified as a priority by more than one ISOC-HAB workgroup,
- deemed necessary as a first step to meet a sequence of research goals,
- addressed issues that were national in scale and scope,
- showed little progress over recent years,
- judged as unlikely to occur without Federal involvement, and
- supported by multiple members of the IWG-4H subcommittee.

Progress toward each research goal will be an important advancement for the respective research priority. Therefore, other than identifying the immediate goals, research and infrastructure goals are not ranked by order of importance. Some research goals are represented under more than one priority, thus, signifying areas where efforts will result in the most rapid advancement. Different Federal agencies have different missions and interests (Table 4), so research goals would logically be approached simultaneously to some degree rather than stepwise. In fact, recent research has made progress toward some goals (see Sections 3.1-3.6), and several agencies are already planning future activities that will further progress in multiple priority areas (see Section 3.7).

CyanoHAB in Siltcoos Lake, Oregon in September 2007. *Photo: Stephen Hager*

4.2. RESEARCH AND INFRASTRUCTURE PRIORITIES

4.2.1. Priority: Improve Methods for Detecting HAB Cells and Toxins

Tools and methods to reliably detect HAB cells in the field and HAB toxins in a variety of matrices (dissolved in water, inside algal cells, and in animal tissues) emerged as the highest priority action, which feeds into goals across all research themes. These tools and methods are a prerequisite for establishing occurrence and prevalence of HABs and their toxins. The ability to *quantify* HAB cells and toxins enables the exploration of health effects, dose-response relationships, ecotoxicology, effective treatment technologies, and establishing the effectiveness of prevention and mitigation strategies. Moreover, reliable detection methods would enable EPA to add freshwater harmful algae and their toxins to the list of agents monitored through the UCM Program. This priority complements and expands for inland waters one of the near-term priorities of the JSOST report, *Charting the Course for Ocean Science in the United States for the Next Decade: An Ocean Research Priorities Plan and Implementation Strategy*, which is to develop new sensors that will, among other things, provide essential information to enable forecasting of ocean-related risks, including those associated with HABs, to human health and safety.

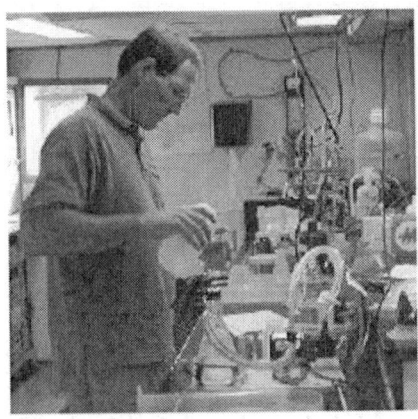

Filtering for microcystin analysis in Lake Erie. *Photo: Steve Wilhelm, University of Tennessee*

Table 3. Research and infrastructure priorities and goals to advance freshwater HAB research and response

Priority	Goals
Improve methods for detecting HAB cells and toxins	• **Develop quick screening methods**[*] • Develop standard methods for analysis • Develop methods for toxins in complex matrices • Identify bioindicators of toxin exposure and effects
Improve understanding of toxin uptake, metabolism, and health effects in humans and other animals	• **Make purified toxins**[*] **available** • Research toxin synthesis • Research effects of toxins and toxin mixtures, including chronic, low-level exposures • Research toxin transfer in foodwebs
Improve human health and ecological risk assessments	• Conduct epidemiological studies to characterize dose-response and identify susceptible populations • Identify bioindicators of toxin exposure and effects • Conduct toxicology studies for specific toxins of concern, for environmentally relevant exposure routes, and for effects of chronic exposures; investigate sub-lethal effects on key aquatic biota • Research toxin fate in aquatic environments and transfer through foodwebs • Consider natural exposures to cyanotoxins occurring in sequence and together for risk assessments
Improve knowledge of bloom occurrence better monitoring	• **Identify and monitor at-risk**[*] • Monitor implicated foods and supplements to identify those at-risk for cyanotoxin contamination • Develop automated detection methods for HABs; couple with observing systems

Table 3. (Continued)

Priority	Goals
	• Determine temporal and spatial trends through long-term monitoring • Collect and store data on bloom characteristics, environmental conditions, and health effects data during blooms
Improve bloom prediction	• Research factors governing algal growth and toxin production • Conduct retrospective analysis to investigate role of human activities • Conduct long-term ecosystem studies on causes and dynamics • Develop predictive ecosystem models for freshwater HABs
Develop HAB prevention control methods	• Develop HAB prevention strategies based on knowledge of causes and • Develop HAB control methods with minimal environmental impacts • Improve drinking water treatment technologies • Conduct cost-benefit analysis for implementation of these strategies
Improve HAB research and response infrastructure	• **Improve availability of toxin standards and other reference materials** [*] • Provide facilities for algal toxin and cell identification • Develop a national database for freshwater HABs • Develop a national-level monitoring strategy • Cultivate taxonomic and toxin expertise for HABs • Educate public to safeguard against exposure and minimize impacts on humans • Improve coordination among programs that address HABs

[*] Bolded goals are considered immediate goals that are critical steps to advancing freshwater HAB research and response.

- **Immediate Goal: Develop quick and reliable screening methods (for cells and toxins) for use in the field. Both manual methods and automated instruments that can be deployed** *in situ* **are important.** Rapid tests will allow responders to determine if toxins are present at levels requiring management action. A tiered approach to toxin identification and quantification will increase response efficiency (e.g., using sentinels or simple tests as the first tier, which would then be corroborated by lab analyses). Automated, *in situ* sensors for specifically and accurately detecting cells or toxins, which can be deployed in the natural environment and relay data back to shore-based facilities in real-time, should be coupled with ongoing and developing observing systems (see monitoring goals, Section 4.2.4). Methods for identification and quantification of HAB cells should be based on a consensus algal taxonomy involving a combination of molecular approaches with microscopic identification.

- **Develop accurate standardized methods for sampling, extracting, identifying, and quantifying toxins.** Methods that can be used by most analytical laboratories using commonly available, affordable instruments are desirable. Appropriate quality assurance/quality control (QA/QC) protocols will ensure comparable results from different laboratories. In the longer term, method development should include protocols for identifying unknown toxins in raw water and extracts from cells and tissues. This is important for determining unknown cyanotoxins as well the toxins of newly emerging freshwater HABs, such as *Prymnesium parvum*.

- **Develop methods for sample preparation and analysis of toxins/metabolites in complex matrices.** As analytical methods improve, it will be possible to determine mixtures of toxins and metabolites in matrices such as biological tissues and sediment.

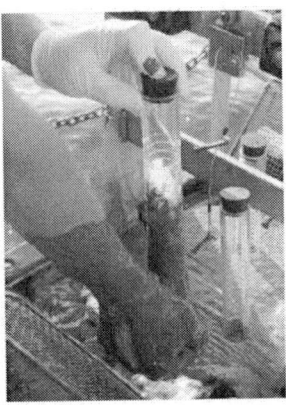

Sediment sampling in Lake Erie. *Photo: Greg Boyer, SUNY ESF*

- **Identify indicators of toxin exposure and effects in animals and humans** (see examples in Section 4.2.3).

Table 4. Agency Interests Organized by Research Priority. Agencies Listed are Those on the IWG-4H and/or those with an Interest in Freshwater HABs as Indicated by Their Participation in ISOC-HAB

Priority	USACE	CDC	EPA	FDA	NASA	NIEHS	NOAA	NSF	USAMRIID	USFWS	USGS	USDA
Improve methods for detecting HAB cells and toxins		X	X	X	X	X		X		X		
Improve understandding of toxin uptake, metabolism, and health effects in humans and other Animals	X	X	X		X	X	X	X		X		
Improve human health and ecological risk assessments		X	X	X		X	X			X	X	
Improve knowledge of bloom occurrence through better monitoring	X	X	X	X	X		X	X		X	X	X

Table 4. (Continued)

Priority	USACE	CDC	EPA	FDA	NASA	NIEHS	NOAA	NSF	USAMRIID	USFWS	USGS	USDA
Improve bloom prediction		X	X		X		X	X			X	
Develop HAB prevention and control methods	X		X				X					X
Improve HAB research and Response infrastructure	X	X	X	X	X	X	X	X	X	X	X	X

4.2.2. Priority: Improve Understanding of HAB Toxin Uptake, Metabolism, and Health Effects in Humans and Animals

Of the freshwater HAB toxins, cyanotoxins currently present the most significant threat to human and ecological health in the United States, so the following research goals are weighted toward cyanotoxins. However, other freshwater HAB toxins, such as the unknown fish-killing toxin(s) produced by *Prymnesium parvum*, can also cause serious ecological and socioeconomic damages, so research goals should also be applied to the study of non-cyanobacteria produced toxins, as appropriate.

Toxin Characterization: Some cyanobacterial toxins have been identified, and their mode of action has been determined; others are partially characterized. There are many variants of cyanobacterial toxins (at least 80 for microcystin), and new toxin analogues continue to be discovered. Toxins are less well characterized for other (non-cyanobacterial) freshwater HABs. Pathways of toxin biosynthesis, and the genes involved, for some CyanoHAB toxins (e.g., microcystins) have been well studied, but the factors that regulate toxin gene expression are not well understood (Neilan et al. 2008).

Toxicokinetics: *Toxicokinetic* research (how toxins are taken up, metabolized, transported, eliminated by an organism) on cyanotoxins to date has been limited to initial efforts to characterize primary metabolism of the more prevalent toxins and basic studies of tissue distribution and elimination. Determinations of the relative contribution of different exposure routes to total exposure, the role of metabolism in toxic responses and detoxification, and particularly, the characteristics of human toxicokinetics, will be important endeavors for future research.

Toxicodynamics: *Toxicodynamic* research (how toxins affect the organism) will help to improve human health and ecological risk assessments. Acute effects of major HAB toxins on humans have been well documented. Less is known about the effects of chronic or repeated, low-level exposures, even though this is likely the primary mode of human and animal exposures. Further, little is known about the effects of less-studied irritant toxins and other cyanobacteria-produced bioactive compounds. Toxins generally occur in mixtures rather than alone, so understanding toxin interactions is also critical (Humpage 2008). Also see Section 4.2.3 for more priorities related to toxins research.

- **Immediate Goal: Purify or synthesize toxins in sufficient quantities for research.** This goal must be accomplished to perform large-scale studies to establish the ecological and human health impacts of HAB toxins (see Section 4.2.3). Studies of the stability and degradation of toxins under laboratory and natural conditions are also required before large-scale studies can be conducted.

- **Research metabolic pathways and genes involved in toxin synthesis.** It is necessary to understand toxin synthesis pathways and the genes that regulate critical steps in order to determine how environmental conditions regulate toxin production (also see Section 4.2.5).

- **Research effects of individual toxins and mixtures (also see Section 4.2.3).** Research would include classical toxicokinetics studies in laboratory animals using individual cyanotoxins and cyanotoxin mixtures often observed in the environment. It will also be important to determine toxicodynamics for priority cyanotoxins and toxin mixtures, with particular focus on repeated, chronic, and low-level exposures.

- **Research mechanisms and magnitude of accumulation of toxins in the food web.** Understanding how toxins move through food webs is important for assessing exposures and potential impacts to ecosystems and human health (also see Section 4.2.3).

4.2.3. Priority: Improve Human Health and Ecological Risk Assessments

Many goals of human and ecological effects research are cross-cutting with those of toxin research as they require advancements in toxicokinetics and toxicodynamics (see Section 4.2.2). Case studies and past animal studies demonstrate that algal toxins pose hazards to humans and ecosystems (see discussion in Chapter 2), but the appropriate data for accurate risk assessments is not yet available.

Human Health Effects

Studies to assess the risk of adverse human impacts are currently limited by the low availability of pure toxins (a priority goal for toxins research, Section 4.2.2, and infrastructure, Section 4.2.7), which are needed to perform toxicity and carcinogenicity studies and to quantify dose-response relationships important for the development of toxicity guidelines to protect human health. Tools to better identify and quantify cause-and-effect relationships will advance assessments of human health risks from cyanotoxin exposure.

- **Conduct epidemiological studies to develop human health risk assessments, including assessments of repeated, low dose exposures.**

    - **Characterize dose-response relationships:** Dose-response data are important for derivation of health guidelines for drinking and recreational waters. Use of quantitative structure activity relationships and toxicity equivalency factor studies are recommended as approaches for filling dose-response data gaps.

- **Identify susceptible populations** using observational studies in areas with documented toxic HAB outbreaks.

- **Identify biomarkers of exposure and effect.** Biomarkers of exposure include the toxin itself, toxin metabolites, and/or toxin-specific altered DNA or proteins (adducts). Biomarkers of effect might include biochemical markers and genotoxicity indicators, such as the number of micronuclei in leukocytes (Hilborn et al. 2008). Biomarkers will strengthen the association between toxin exposure and health effects.

- **Conduct neurodevelopmental and immunotoxicological studies for all groups of cyanobacterial toxins.**

- **Conduct studies to reduce uncertainties in risk assessments for specific toxins.**

  - **Cylindrospermopsin – carcinogenicity studies:** The toxin structure and preliminary studies suggest carcinogenicity of cylindrospermopsin, which requires high priority investigation (Falconer 2005, Hilborn et al. 2008, Donohue et al. 2008).

  - **Microcystin-LR – chronic and reproductive toxicity studies:** Such studies would be the most beneficial to reduce uncertainties in risk assessment for this toxin (Donohue et al. 2008).

  - **Anatoxin-A – sub-chronic and developmental toxicity studies:** Past studies have shown the acute neurotoxicity of anatoxin-A. Thus, studies looking at the health effects of moderate duration exposures would be the next priority to reduce uncertainty of risk assessments (Donohue et al. 2008).

- **Research environmentally relevant exposure routes** (oral, inhalation, and dermal) and the effects of acute, episodic, and chronic exposures at low (sub-lethal) concentrations (also see Section 4.2.2).

Warning posted in Lake Agawam, New York, to notify public of precautions to take during CyanoHAB events to reduce risk of illness. *Photo: Chris Gobler, SUNY*

Ecological Effects

A major uncertainty in assessments of ecological effects from freshwater HABs is the specific role that toxins play in adverse impacts inflicted by blooms. Most research on toxic effects has involved exposures of single species to individual toxins in controlled lab experiments. Moreover, most studies have focused on short-term exposures to toxins even though the exposure of aquatic biota in nature can be intermittently recurring and range from days to years (Fournie et al. 2008). A scaling up of ecological effects research to include entire food webs and longer time periods would offer significant improvement.

- **Research the sub-lethal effects of chronic toxin exposures for key taxa of aquatic biota**, including invertebrates, fish, and others, beginning with in vitro work and proceeding to in vivo studies as appropriate. Studies should include investigation of physiological, pathological, and behavioral effects.

- **Research toxin fate and transformation in the aquatic environment and the transfer through food webs.** Studies to understand the bioaccumulation the bioconcentration, and the biomagnification of different cyanotoxins and other cyanobacterial bioactive compounds in food webs, as well as models that describe the fate of cyanotoxins in water, sediments, and other organisms, will be important for assessing human exposures and health effects. These studies and models will also help clarify the relative importance of the effects of cyanotoxins versus the effects of cyanobacteria at the ecosystem level.

Human and Ecological Risk Assessments

- **Develop risk assessments (for humans and aquatic biota) of naturally occurring toxic bloom events, including those in which various HAB species occur together or in sequence.** Risk assessments of populations exposed to natural blooms need to consider a range of pathological effects generated by different cyanobacterial toxins, in sequence and in combination.

Dead fish in Lake Erie. *Photo: Steve Wilhelm, University of Tennessee*

4.2.4. Priority: Improve Knowledge of Bloom Occurrence Through Better Monitoring

Monitoring for cells and toxins will advance knowledge of bloom occurrence, which is important to understand the prevalence and severity of freshwater HABs and to prevent or reduce their impacts on humans and ecosystems. Effective monitoring can be challenging, however, due to difficulties sampling at adequate temporal and spatial scales and the expense required for sampling, analysis, and testing. The ideal monitoring system would provide real-time, highly automated, accurate bloom detection that in the short run could provide early warning of impending toxic events and in the long run will lead to better predictive capability.

Many research priorities in other categories lead to better and more efficient monitoring. Tools to advance monitoring include easier, cheaper, faster, and more accurate methods for detecting cells and toxins (see Section 4.2.1), citizen monitoring networks (see Section 4.2.7), databases, (see Section 4.2.7) and diagnostic tools for monitoring exposure and effects in humans and higher trophic level sentinel species (see Sections 4.2.1 and 4.2.3). Because these issues are addressed in other sections, the monitoring priorities described below focus on monitoring approaches and implementation.

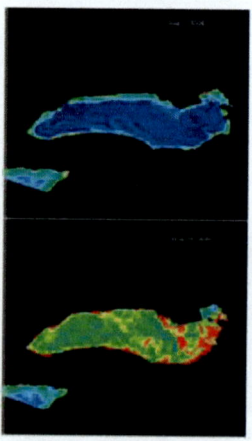

SeaWiFs images of Lake Ontario on 01Aug04 (top) and 16Aug04 (bottom). Images have been processed to show chlorophyll concentration. *Images: MERHAB Lower Great Lakes project http://www.esrs.wmich.edu/merhab.htm*

Some states and localities are already monitoring for HABs in recreational and drinking water (see Appendix III). Some of the research described below will provide states with improved capabilities to continue their monitoring. Other research will focus on the national scope of this problem and help to clarify the Federal role in preventing blooms and their impacts.

- **Immediate Goal: Identify and monitor waterbodies used for drinking or recreation that are at risk of for toxin-producing blooms (essential for understanding potential health effects from exposure to toxin mixtures).** A tiered approach to monitoring in which field-based monitoring is corroborated by laboratory analysis would be beneficial. A nationwide survey for the occurrence of HAB cells and toxins in drinking water and recreational water would be helpful to determine 'at risk' waterbodies. Surveillance for newly emerging toxins is also important.

- **Monitor implicated foods and nutritional supplements to identify products at risk for contamination with cyanobacterial toxins.**

- **Develop methods for *in situ* detection of cells and toxins using automated instruments, and for remote detection of HABs using satellites. Couple HAB monitoring with ongoing and developing observational programs (e.g. Integrated Ocean Observing System, or IOOS; Integrated Earth Observing System, or IEOS; Coastal Global Ocean Observing System, or GOOS)** (also see Section 4.2.1).

- **Support long term monitoring to determine temporal and spatial occurrence of HAB cells and toxins.**

- **Collect and store data on bloom characteristics and the environmental conditions that exist during the blooms as well as human and animal health effects, with the intent of developing models to predict bloom occurrence and impacts.**

4.2.5. Priority: Improve Bloom Prediction

An important goal of much freshwater HAB research is to sufficiently understand the environmental and biological factors controlling HAB dynamics so models can be developed to predict the occurrence and toxicity of HABs, bloom transport, and the fate of toxins. Developing predictive models requires a thorough understanding of the interaction of physiological, ecological, and limnological factors. By identifying critical factors regulating bloom dynamics, these models can also help refine management strategies to reduce or prevent HABs (also see Section 4.2.6).

- **Research the environmental factors governing algal growth and toxin production.** Laboratory cultures and microcosm/mesocosm studies on natural populations can be used to measure growth and toxin production of multiple species and strains in relation to a variety of environmental variables (e.g. nutrients, hydrology, temperature, light, carbon dioxide, viruses, bacteria, other algae, grazers, and higher trophic levels). In addition, exploring environmental factors regulating toxin production at the genetic level will be a valuable tool toward reaching this goal (also see Section 4.2.2).

- **Conduct retrospective analyses to examine relationships between human activities and regional expansion of freshwater HABs along temperature, precipitation, and nutrient gradients.** Available data from existing literature, long-term observational programs, proxy indicators in sediment cores, and data sets of environmental patterns should be correlated with cyanobacterial abundance where possible.

- **Conduct long-term studies with an ecosystem approach to determine the causes and dynamics of HABs in the natural environment.**

- **Develop predictive ecosystems models for freshwater HABs.** These would require the information and understanding achieved from all the research described above, uniting laboratory and field data, including data from ongoing monitoring programs (see Section 4.2.4), to develop models that can predict HABs and be used to assess strategies for preventing HABs (see Section 4.2.6).

4.2.6. Priority: Develop HAB Prevention and Control Methods

The definitive solution for managing HABs would be to avoid the problem altogether through preventive strategies, but because HABs are a complex natural phenomenon, prevention may not be feasible in many cases. Prevention is attainable in some instances, however, since some HABs are influenced or exacerbated by human activities that can be managed (see Box 2.2). Developing better prevention strategies requires a thorough understanding of the biological and environmental factors regulating bloom dynamics. Controlling HABs (i.e., reduction or containment of an existing bloom) is another desirable management strategy, but can also be challenging due to costs, limits of effectiveness, potential environmental impacts, and negative public perceptions. Treatment to remove toxins is the primary management strategy employed in drinking water facilities, but these strategies are expensive and still in the early stages of development. Effective treatment for removal of toxins from drinking water will be essential if a regulatory process is put in place. Methods for managing toxins in recreational waters or in freshwater fish lag behind those for drinking water because the impacts are not as well understood and treatment in large basins would be more difficult. Cost-benefit analysis is useful for evaluating the overall value of management approaches that implement prevention and control strategies.

CyanoHAB in Lake Neatahwanta, New York. *Photo: Greg Boyer*

CyanoHAB in Hamilton Harbor, Lake Ontario. *Photo: Greg Boyer, SUNY ESF*

- **Develop HAB prevention strategies using information on the causes of freshwater HABs.** In some cases, reducing nutrient loads to surface waters is known to prevent or reduce HABs. Also, management practices that change nutrient ratios may alter bloom species composition to less toxic species. Alterations in hydrology (such as increased flushing) may be an effective strategy in some locations if upstream water supplies are available. Models based on understanding of HAB dynamics can be used to test such prevention strategies.

- **Develop effective HAB control methods that have minimal impacts on the environment.** Potential control techniques to investigate further include artificial destratification, increasing flushing rates, ultrasound, electrocoagulation, new and existing coagulants, and new algicidal or algistatic compounds.

- **Develop effective treatment technologies to remove cyanotoxins from drinking water.** Investigations of enhanced coagulation technology, filtration effectiveness, and disinfectant by-products are important. Microcystins, cylindrospermopsin, and anatoxin-A are the primary algal toxins of concern for regulation under the SDWA.

- **Conduct cost-benefit analyses for management strategies.** Estimates of the cost for prevention and control strategies compared to the cost of blooms to society through impacts on food, drinking water, recreation, natural resources, as well as aesthetic impacts and lost ecosystem services will be beneficial for developing cost-effective management strategies.

4.2.7. Priority: Improve HAB Research and Response Infrastructure

Different types of infrastructure support the ability to monitor, predict, mitigate, control, and prevent HABs. Goals for strengthening infrastructure to support all HAB research and response have been highlighted in the HARRNESS (2005) report and in the *National Assessment of Efforts to Predict and Respond to Harmful Algal Blooms in U.S. Waters* (Jewett et al. 2007). Infrastructure includes standards and probes, shared-use facilities, databases, platforms for real-time monitoring, and training to develop expertise. Improvements in freshwater HAB infrastructure will support state-of-the-art HAB monitoring and detection and lead to more accurate risk assessments and HAB predictions.

- **Immediate Goal: Improve availability of toxin standards and other reference materials.** Reference materials, including certified toxin standards, are used for method development, confirming identification, developing new probes and assays, training, generating reliable quantitative data on toxins, and determining toxicological properties of specific toxins. These materials need to be widely available in sufficient quantities for research. In addition, standardized methods that can be used by most analytical laboratories or used in the field are critical.

  - **Improve availability of both certified toxin standards and toxins labeled with either stable or radioisotopes.** In particular, reference standards for microcystin-LR, anatoxin A, and cylindrospermopsin are critical (also see immediate goal in Section 4.2.2).

- **Maintain and improve culture collections that provide a variety of toxic and nontoxic freshwater algae.** In the absence of purified toxins, bulk culturing of toxigenic strains can yield much needed material for toxicology experiments. Further, the availability of cultures is critical for training in cell identification and conducting culture studies of HAB growth and toxin production under controlled conditions.

- **Provide facilities for toxin and cell identification.**

  - **Establish shared facilities for cell identification and toxin analysis, including both routine and novel toxins.**

  - **Develop, standardize, and make available sample collection and preparation techniques and analytical methods for toxins** (also see Section 4.2.1). Details of issues to consider for standardizing methods are reviewed by de la Cruz et al. (2008).

Develop a national database for freshwater HABs. Databases serve as a means of collecting data from disparate sources into one readily accessible location. Well designed databases can prevent duplication of effort and allow data collected for one purpose to be used for other related purposes.

- **Develop a national database for CyanoHABs and other freshwater HABs that standardizes taxonomy to aid managers and researchers in HAB identification.** Several databases developed for the Baltic Sea illustrate the types of information that databases like this might include (e.g., http://www.helcom.fi/groups/monas/en_GB/biovolumes/and http://www.smhi.se/oceanografi/oce_info_data/plankton_checklist/ssshome.htm). A database of this type would support training and education/outreach activities and promote consistency among researchers.

- **Compile existing information on HAB occurrence into an accessible form that could serve as a routinely updated, national database** (e.g., by expanding HABISS). Such a database could also support retrospective analysis of HAB occurrence in order to examine relationships between human activities and changes in HAB distributions and occurrence.

- **Integrate U.S. databases with an international database, such as CYANONET** (see Section 5.2.1). An integration of databases with different types of information (e.g. HAB occurrence, environmental data, health and ecological effects, and research project results) would promote better collaboration and reduce duplication. Moreover, an internationally integrated database for occurrence, regulations, guidelines, management strategies, and outcomes as described by various countries, states, and local governments would accelerate efforts to develop scientifically-sound management strategies.

- **Develop a national-level monitoring strategy.** Monitoring programs for detecting and validating HAB events allow for a better understanding of freshwater HABs and their impacts. Tools and approaches to improve monitoring programs are addressed Sections 4.2.1 and 4.2.4. Infrastructure that would enable desirable monitoring approaches is described below.

  - **Forge a national-level agreement on monitoring strategy,** including Federal guidelines to determine when beach closings and health advisories are needed.
  - **Design and integrate site observing systems for monitoring HABs in areas with frequent blooms.** This goal involves developing and integrating automated, *in situ* HAB-specific sensors.
  - **Continue the development of HABISS as a national freshwater HAB surveillance system with a centralized database** (see Box 3.8).

- **Cultivate taxonomic and toxin expertise.** Training involves developing expertise within the freshwater HAB management and research communities for HAB species and toxin identification. Cultivation of taxonomic and toxin expertise is essential as the frequency and extent of known HABs increase and new species and toxins are identified (especially since fewer people are choosing to become experts in HAB identification). Such training will be beneficial at a wide range of levels, from citizen monitoring groups and local resource managers in impacted regions to researchers who want to specialize in HAB taxonomy or toxin analysis.

- **Provide training in HAB cell identification and toxin analysis at a variety of levels.** Taxonomic training should include both microscopic and molecular identification methods and be widely available for a range of expertise, including state, local, and tribal personnel. In addition, training citizens in HAB identification will contribute to the success of monitoring programs that use citizen monitoring networks as part of a tiered monitoring approach.

- **Educate public to safeguard against exposure and to minimize impacts.** Outreach promotes community awareness of freshwater HAB issues. The *HARRNESS* report (HARRNESS 2005) and the Harmful Algal Research and Response: A Human Dimensions Strategy report (HARR-HD 2006) emphasize the importance of education and outreach to populations most susceptible to HAB impacts. Outreach can lessen HAB impacts by promoting awareness of potential threats, by imparting accurate perceptions of drinking water, recreational water, and seafood (freshwater fish and crustaceans) safety within the community, and by fostering community participation in HAB prediction and response efforts. Citizen monitoring networks are an example of an outreach/training activity that benefits local communities as well as the broader management community.

- **Develop outreach programs to educate the public about the causes and risks of freshwater HABs and to inform the public** (especially stakeholders most affected) as freshwater HAB knowledge and management strategies evolve.

- **Improve coordination among programs that address HABs** (also see Chapter 5). Better coordination of existing resources at the national and international levels will improve efficiency of freshwater HAB research and management.

- **Improve communication and coordination between Federal agencies and other entities participating in freshwater research and management,** including state agencies (see Appendix III), local governments, and non-profit organizations (see Appendix II).

- **Improve event response for freshwater HABs.** Currently there is no national freshwater program for event response analogous to those for marine systems (such as the NOAA CSCOR HAB Event Response Program and the NOAA Marine Biotoxins Program's Analytical Response Team). The *RDDTT Plan* (Box 1.1) will address ways to improve Event Response programs in a general sense. Some state health agencies (e.g., Florida, North Carolina, Virginia, Maryland, and South Carolina) have developed their own response strategies (see Appendix III).

- **Improve international communication and coordination.** A number of other countries have had to confront the problems associated with freshwater HABs much sooner and comprehensively than the United States, so U.S. research and management can benefit from lessons learned elsewhere.

Anglers fishing on Lost Creek in Oregon. *Photo: USACE*

In: Marine and Freshwater Harmful Algal Blooms   ISBN: 978-1-60741-838-2
Editor: Peter E. Williams   © 2010 Nova Science Publishers, Inc.

Chapter 5

STEPS TO IMPROVE COORDINATION AND COMMUNICATION FOR FRESHWATER HAB RESEARCH AND RESPONSE

Joint Subcommittee on Ocean Science and Technology

Plans for improving coordination and communication in response to the growing problem of freshwater HABs in the United States require an evaluation of interactions and information exchange at the national and the international levels. Recommendations for improving the flow of information and the coordination of plans are described in this chapter. This effort recognizes the need for increased interaction between agencies, partners, and stakeholders in order to optimize freshwater HAB prevention, prediction, and response.

Improving coordination and communication will lead to improvements in: 1) interagency intramural research program planning, 2) interagency extramural research program planning, 3) intramural and extramural program planning integration, 4) Federal, state, local, private, academic, and international research program planning implementation, and 5) dissemination of research results. These improvements will also speed and enhance the development of infrastructure, tools, information, and guidance needed by state, local, and tribal governments to develop options and strategies for reducing the risks posed by HABs and HAB toxins in freshwater recreational and drinking water sources. Improved communication and coordination will help prevent duplication of effort.

5.1. NATIONAL COORDINATION AND COMMUNICATION

Improving coordination and communication among Federal agencies and between Federal agencies and other levels of government and organizations with a role in freshwater HAB research or response has four aspects: 1) devising an ongoing mechanism for communication and coordination between Federal agencies, 2) increasing representation of freshwater HAB researchers and managers in HAB community coordination, 3) promoting involvement in national meetings and 4) recognizing the important role of the U.S. National Office for Harmful Algal Blooms.

5.1.1. Federal Coordination

The 2004 reauthorization of HABHRCA calls for this scientific assessment to "identify ways to improve coordination and to prevent unnecessary duplication of effort among Federal agencies and departments with respect to research on HABs in freshwater locations" (Public Law 108-456). HABHRCA 2004 also calls for the development of three other reports about different aspects of HAB research and response (see Box 1.1), all of which are required by the legislation to address Federal coordination.

Federal coordination for HAB research and response, including freshwater HABs, should be provided by the IWG-4H. As described in Chapter 1, the IWG-4H was formed by the JSOST to fulfill the requirement of HABHRCA 2004 to reestablish the Interagency Task Force on HABs and Hypoxia (see Box 1.2). Section E.2 of the Charter of the IWG-4H specifies that it will "Ensure interagency communication, coordination and cooperation."

5.1.2. Coordination within the HAB Research and Management Community

The *HARRNESS* report (HARRNESS 2005) called for the formation of a National HAB Committee (NHC), whose purpose is to facilitate coordination and communication of activities for the U.S. HAB community at a national level. The NHC has recently been constituted and is in the process of developing its terms of reference and procedures. The NHC is an elected body with members representing the HAB research and state and local management community with non-voting *ex officio* members from some Federal agencies. The IWG-4H and NHC would ideally be linked by the Federal *ex officio* members that serve on the NHC (see Box 5.1) so that common coordination activities could be undertaken. In order to serve the freshwater HAB research and management communities as well as the marine HAB community, the following modifications of the NHC would be beneficial:

1. The membership of the NHC might be broadened to reflect balance with regard to freshwater and marine interests.
2. *Ex officio* members might be appointed to represent the agencies and non-profit organizations (e.g., AwwaRF) that sponsor or conduct research on freshwater HABs.
3. A freshwater subcommittee might be formed to develop and foster coordination activities similar to those that are already on-going in the marine HAB community.
4. An interprogram coordination subcommittee might be formed to develop the framework and strategy for communication and coordination, with multiple national and international programs which are relevant to HAB issues, including freshwater HABs.
5. An infrastructure subcommittee might be formed to develop a strategy for implementation of the infrastructure goals identified in the HARRNESS report and in this report.

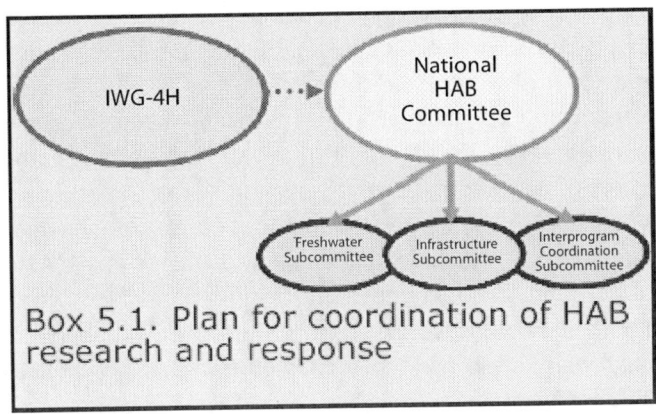

Box 5.1. Plan for coordination of HAB research and response

> **BOX 5.2. NATIONAL HAB MEETINGS**
>
> - The U.S. HAB symposia series is a biennial series focusing on HAB issues in the United States. The fourth meeting was held October 29-November 2, 2007 in Woods Hole, Massachusetts.
> - The Gordon Research Conference on Mycotoxins and Phycotoxins provides an international forum for the presentation and discussion of leading edge research on the biology, chemistry, risk assessment and toxicology of fungal, algal, and cyanobacterial toxins that are known food or water contaminants. The most recent meeting was held June 17-22, 2007, at Colby College in Waterville, Maine. http://www.grc.org/

5.1.3. National Meetings and Scientific Societies

Currently in the United States, communication between HAB researchers and Federal, state, and local managers is facilitated by several national meetings that occur on a regular basis and are partly sponsored by Federal agencies. The latest research findings and their application to resource and public health management are the main topics of these meetings. These include the biennial U.S. HAB Symposium and the biennial Gordon Research Conference on Mycotoxins and Phycotoxins (see Box 5.2). In recent years both meetings have included freshwater HABs among the topics discussed, but given the increasing severity of the problem, it should receive more emphasis. One question that could be addressed by the proposed NHC sub-committee on freshwater HABs is whether separate meetings are needed to adequately cover freshwater HAB issues, or if the current structure can be expanded to provide better coverage of freshwater HAB issues.

Other organizations—such as the American Water Works Association (http://www.awwa.org/), the North American Lake Management Society (www.nalms.org), the International Association for Great Lakes Research (http://www.iaglr.org/), and the Aquatic Plant Management Society (http://www.apms.org) —include management of freshwater HABs at regional and national meetings. For example, the results of a 2005 North American Lake Management Society Special Symposium on Cyanobacterial HABs in Madison, Wisconsin is available as a CD set at *http://www.nalms.org/* News/CyanoDVDs.aspx.

5.1.4. Role of the U.S. National Office for Harmful Algal Blooms

The U.S. National Office for Harmful Algal Blooms, established with funding from NOAA CSCOR, provides critical infrastructure, including communication, coordination, and technical support capabilities that enhance the Nation's ability to respond to and manage the growing threat posed by HABs. This office has recently expanded its scope to include freshwater HAB issues. It aids coordination by organizing HAB meetings, symposia, and workshops and facilitates training and student participation in HAB activities. Communications are facilitated by maintaining the Harmful Algae Page website (http://www.whoi.edu/redtide/), hosting list servers for national and regional HAB issues, archiving HAB reports, and providing a central location for announcements of funding opportunities and meetings. The National Office for Harmful Algal Blooms also maintains databases of U.S. HAB events that are coordinated at the international level, provides technical information through its web site, and assists with the preparation of national and international HAB reports. The offices's activities also facilitate the functions of the NHC.

5.2. INTERNATIONAL COORDINATION AND COMMUNICATION

Freshwater HABs are a global problem (Carmichael 2008) and some countries have well-established research programs and response plans. Better coordination and communication at the international level will help U.S. efforts advance at a rapid pace and avoid duplication of effort.

BOX 5.3. INTERNATIONAL ORGANIZATIONS WITH INTERESTS IN HAB RESEARCH AND RESPONSE

Blue represents organizations focused primarily on marine HABs and green indicates organizations with interests on freshwater HABs. For more information about these organizations, go to the following web sites CYANONEThttp://www.cyanonet.org,GEOHAB   http://ioc.unesco.org/hab/GEOHAB.htm IOC HAB Program

---

http://ioc.unesco.org/hab/intro.htm SCOR http://www.jhu.edu/~scor/, IOC http://ioc.unesco.org/iocweb/index.php,IHP http://typo38.unesco.org/index.php?id=240, ICSU http://www.ICSU.org, UNESCO *http://portal.unesco.org*

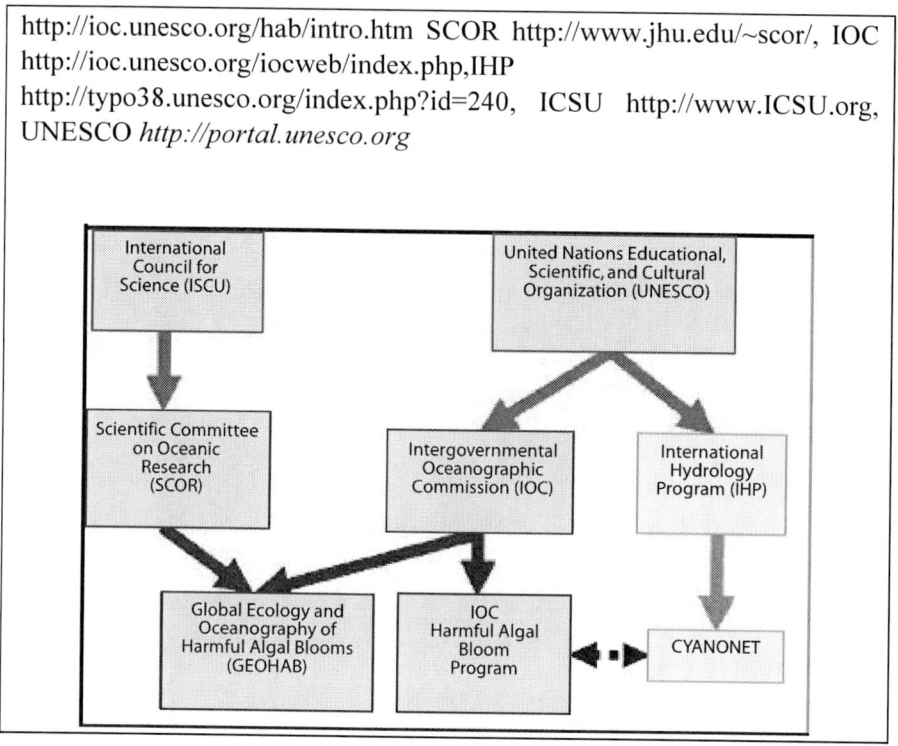

5.2.1. U.S. Participation in International Organizations

CYANONET (see Box 5.3) is a global network for the management of CyanoHABs and toxins in water resources that was established as part of the United Nations Educational, Scientific, and Cultural Organization's (UNESCO) International Hydrology Program. Its initial goal is to carry out a global assessment of the occurrence, health, and economic significance of cyanobacteria and cyanotoxins. In the longer term, it aims to increase awareness of the problem and potential solutions by the dissemination of information, training, and education and by sharing experience and best management practices. U.S. participation in global planning efforts has been limited, although AwwaRF, EPA, and some academics have been important contributors from the United States in these efforts. CYANONET provides an excellent opportunity to expand U.S. support and cooperation in freshwater CyanoHAB research and response and could be used as a platform for building and the integrating of a United States database. U.S. participation would increase opportunities to improve U.S. freshwater HAB research and response capabilities.

The Association of Analytical Communities (AOAC) International's Marine and Freshwater Toxins Task Force (http://www.aoac.org/ marine_toxins/task_force.htm#Overview) is an international group of experts on marine and freshwater toxins and stakeholders who have a strong and practical interest in the development and validation of methods for detection of these toxins. In response to the global need for improved testing methods for these toxins, the Task Force validates and provides training in new methods. Regional meetings, online forums, and journals of the AOAC provide an opportunity for exchange of information about new methods.

In addition to the international activities described above, binational coordination between the United States and Canada on water quality issues, including algal blooms, in the Great Lakes is conducted by the International Joint Commission (http://www.ijc.org/en/home/main_accueil.htm). Several countries with significant CyanoHAB problems have their own national CyanoHAB working groups, including Australia (Cooperative Research Center for Water Quality and Treatment, http://www.waterquality.crc.org.au/), Japan (National Institute for Environmental Studies, Tsukuba, *http://www.nies.go.jp/ index.html*), and China (Institute of Hydrobiology, Chinese Academy of Sciences, Wuhan, http://159.226.163.238/ english/default.aspx). Coordination of these groups with an equivalent body in the United States and use of compatible databases integrated with a common database, like CYANONET, would avoid duplication of effort.

5.2.2. International Scientific Meetings

International meetings (see Box 5.4) attended by HAB researchers and managers provide a forum to exchange recent research results that may lead to new management approaches. The International Conference on Toxic Cyanobacteria focuses on cyanobacterial physiology, ecology, molecular biology, toxicology, risk assessment, management, and toxin detection. The International Conference on Harmful Algae covers all HABs, but its focus on freshwater HABs increased substantially at the most recent meeting in 2006. NOAA and NSF, as well as other countries and international organizations, have provided support. The International Union of Pure and Applied Chemistry (IUPAC) International Symposia on Mycotoxins and Phycotoxins (of which cyanotoxins are a subset) have a broader focus on research in toxin analysis, human health effects, risk assessments, and control and treatment strategies. U.S. participation in and support of these meetings benefit both the United States and the international community.

Communication between the United States and Canada concerning HABs in the Great Lakes is facilitated by a number of binational meetings, including the EPA-sponsored biannual State of the Lakes Ecosystem Conference, (http://www.epa.gov/glnpo/solec/) and meetings of the International Association for Great Lakes Research (http://www.iaglr.org/).

BOX 5.4. INTERNATIONAL HAB MEETINGS

- International Conference on Toxic Cyanobacteria is held every 3 years. The most recent conference was held Brazil in 2007. *http://www.biof.ufrj.br/cyano/main.html*
- International Conference on Harmful Algae is held every 2 years. The next one (13th) will be in Hong Kong in 2008. http://www.hab2008.hk/
- International Symposium on Mycotoxins and Phycotoxins. The most recent one (12th) was held in Turkey in May 2007. http://iupac2007-mycotoxin.atal.tubitak.gov.tr/

In: Marine and Freshwater Harmful Algal Blooms   ISBN: 978-1-60741-838-2
Editor: Peter E. Williams   © 2010 Nova Science Publishers, Inc.

REFERENCES

Anadotter, H., Cronberg, G., Lawton, L. A., Hansson, H. B., Gothe, U. & Skulberg, O. M. (2001). An extensive outbreak of gastroenteritis associated with the toxic cyanobacterium *Planktothrix agardii* (Oscillatoriales, Cyanophyceae) in Scania, South Sweden, In: I. Chorus, (Ed.) *Cyanotoxins, occurrence, causes, consequences.* Berlin, Springer, 200-208.

Atech. (2000). Cost of algal blooms. *Report to Land and Water Resources Research and Development Corporation*, Canberra, ACT 2601 ISBN 0 *642*, 76014 4.

Backer, L. C. (2002). Cyanobacterial harmful algal blooms (CyanoHABs): Developing a public health response. *Lake and Reservoir Management, 18(1)*, 20-31.

Barrett, P. R. F., Littlejohn, J. W. & Curnow, J. (1999). Long-term algal control in a reservoir using barley straw. *Hydrobiologia. 415*, 309-313.

Bootsma, H. A., Jensen, E. R., Young, E. B. & Berges, J. A. (2004). *Cladophora* research and management in the Great Lakes. *Proceedings of a Workshop Held at the Great Lakes WATER Institute, University of Wisconsin, Milwaukee*, December 2004.
http://www.uwm.edu/Dept/GLWI/cladophora/page_report.html.

Boyer, G. L. Watzin, M. C., Shambaugh, A. D., Satchwell, M. F., Rosen, B. H. & Mihuc, T. (2004). The occurrence of cyanobacterial toxins in Lake Champlain. In: G. L., Boyer, M. C., Watzin, A. D., Shambaugh, M. F. Satchwell, B. H. Rosen, & T. Mihuc., (Eds.): *Lake Champlain: Partnerships and Research in the New Millennium*, New York: Kluwer Academic/ Plenum Publ. 241-257.

Boyer, G. L. (2008). Cyanobacterial toxins in New York and the Lower Great Lakes ecosystems. In: H. K. Hudnell, (Ed.) *Cyanobacterial Harmful Algal*

Blooms: State of the Science and Research Needs. Advances in Experimental Medicine & Biology, Vol. 619. Springer. 500.

Burns, J., Williams, C. & Chapman, A. (2002). Cyanobacteria and their toxins in Florida Surface Waters. In: D., Johnson, R. D. Harbison, (Eds.) Proceedings of Health Effects of Exposure to Cyanobacteria Toxins: *State of the Science*, August 13-14, 2002, 16-21.

Burns, J. (2008). Toxic cyanobacteria in Florida waters. In: H. K. Hudnell, (Ed.) *Cyanobacterial Harmful Algal Blooms: State of the Science and Research Needs. Advances in Experimental Medicine & Biology*, Vol. *619*. Springer. 500.

Carmichael, W. W. (1992). A status report on plankton cyanobacteria (blue-green algae) and their toxins. *Office of Research and Development*, EPA: Washington, D.C. Report # EPA/600/R-92/079. 141.

Carmichael, W. W. (2001). Assessment of blue-green algal toxins in raw and finished drinking water. Denver: AWWA *Research Foundation and American Water Works Association*. ISBN 1-58321-076-8. 1-49.

Carmichael, W. W. (2008). A world overview one-hundred, twenty-seven years of research on toxic cyanobacteria--Where do we go from here? In: H. K. Hudnell, (Ed.) *Cyanobacterial Harmful Algal Blooms: State of the Science and Research Needs. Advances in Experimental Medicine & Biology*, Vol. 619. Springer. 500.

Carmichael, W. W., Azevedo, S. M. F. O., An, J. S., Molica, R. J. R., Jochimsen, E. M., Lau, S., Rinehart, K. L., Shaw, G. R. & Eaglesham, G. K. (2001). Human fatalities from cyanobacteria: Chemical and biological evidence for cyanotoxins. *Environmental Health Perspectives, 109(7)*, 663-668.

Chapman, A. D. & Schelske, C. L. (1997). Recent appearance of Cylindrospermopsis (Cyanobacteria) in five hypereutrophic Florida lakes. *J. Phycol, 33*, 191-195.

Codd, G. A., Azevedo, S. M. F. O., Bagchi, S. N., Burch, M. D., Carmichael, W. W., Harding, W. R., Kaya, K. & Utkilen, H. C. (2005). CyanoNet: A global network for cyanobacterial bloom and toxin risk management. International Hydrological Programme. Initial situation assessment and recommendations. *Technical Documents in Hydrology*, No. 76, 138 Unesco, Paris. http :// unesdoc. Unesco .org/ images /0014/001425/ 142557E.pdf

Cox, P. A., Banack, S. A. & Murch, S. J. (2003). Biomagnification of cyanobacterial neurotoxins and neurodegenerative disease among the

Chamorro people of Guam. *Proceedings of the National Academy of Sciences, 100(23)*, 13380-13383.

Cox, P. A., Banack, S. A., Murch, S. J., Rasmussen, U., Tien, G., Bidigare, R. R., Metcalf, J. S., Morrison, L. F., Codd, G. A. & Bergman, B. (2005). Diverse taxa of cyanobacteria produce β-N-methylamino-L-alanine, a neurotoxic amino acid. *Proceedings of the National Academy of Sciences. 102(14)*, 5074-5078.

de la Cruz, A. A., Meyer, M. T., Echols, K., Furey, A., Hungerford, J. M., Lawton, L., Mandeville, R., Meriluoto, J. A. O., Rublee, P., Sivonen, K., Stelma, G., Wilhelm, S. & Zimba, P. A. (2008). Analytical methods workgroup report. In: H. K. Hudnell, (Ed.) Cyanobacterial Harmful Algal Blooms: State of the Science and Research Needs. *Advances in Experimental Medicine & Biology*, Vol. *619*. Springer. 500.

Dillenberg, H. O. & Dehnel, M. K. (1960). Toxico waterbloom in Saskatchewan, 1959. *Canadian Medical Association Journal, 83*, 1151-1154.

Donohue, J., Orme-Zavaleta, J., Burch, M., Dietrich, D., Hawkins, B., Munns, W. R.., Steevens, J., Steffensen, D., Stone, D. & Tango, P. (2008). Risk assessment workgroup report. In: H. K. Hudnell, (Ed.) Cyanobacterial Harmful Algal Blooms: State of the Science and Research Needs. *Advances in Experimental Medicine & Biology. Vol. 619*, Springer. 500.

Dortch, Q., Peterson, T. D., Achee, S. & Furr, K. L. (2001). Phytoplankton, cyanobacterial blooms, and N_2 Fixation in Years with and without Mississippi River Diversions. In R. E., Turner, D., Justic, Q. Dortch, & N. N. Rabalais, (Eds.) Nitrogen Loading into Lake Pontchartrain. *Report to Lake Pontchartrain Basin Foundation*, Metairie, LA.

ECOHAB. (1995). *The ecology and oceanography of harmful algal blooms. A national research agenda*. D. A. Anderson, (Ed.) Woods Hole Oceanographic Institution, Woods Hole, MA. 66.

Edmondson, W. T. & Lehman, J. T. (1981). The effect of changes in the nutrient income on the condition of Lake Washington. Limnol. *Oceanogr. 26(1)*, 1-29.

Falconer, I. R. (1993). *Algal toxins in drinking water. London*, Academic Press. 224.

Falconer, I. R. (2005). *Cyanobacterial toxins of drinking water supplies: cylindrospermopsins and microcystins*. CRC Press. Boca Raton, FL. *279*.

Falconer, I. R. (2008). Health effects associated with controlled exposures to cyanobacterial toxins. In: H. K. Hudnell, (Ed.) Cyanobacterial Harmful

Algal Blooms: State of the Science and Research Needs. *Advances in Experimental Medicine & Biology, Vol. 619.* Springer. 500.

Fournie, J. W., Hilborn, E. D., Codd, G. A., Coveney, M., Dyble, J., Havens, K., Ibelings, B. W., Landsberg, J. & Litaker, W. (2008). Ecosystem effects workgroup report. In: H. K .Hudnell, (Ed.) Cyanobacterial Harmful Algal Blooms: State of the Science and Research Needs. *Advances in Experimental Medicine & Biology, Vol. 619,* Springer. 500.

Fristachi, A., Sinclair, J. L., Hambrook Berkman, J. A., Boyer, G., Burkholder, J. A., Burns, J., Carmichael, W., DuFour, A., Frazier, W., Morton, S. L., O'Brien, E. & Walker, S. (2008). Occurrence of cyanobacterial harmful algal blooms workgroup report. In: H. K. Hudnell, (Ed.) Cyanobacterial Harmful Algal Blooms: State of the Science and Research Needs. *Advances in Experimental Medicine & Biology, Vol. 619,* Springer. 500.

Fujiki, H., Suganuma, M., Suguri, H., Yoshizawa, S., Takagi, K., Nakayasu, M., Ojika, M., Yamada, K., Yasumoto, T., Moore, R. E. & Sugimura, T. (1990). New tumor promoters from marine natural products. In: S. Hall, & G. Strichartz, (Eds.) *Marine Toxins: Origin, Structure and Molecular Pharmacology,* American Chemical Society, Washington D.C., 232-240.

Glass, J. (2003). Historical Review of Golden Alga (*Prymnesium parvum*) Problems in Texas. October 2003 Golden Alga Workshop Abstracts. http://www.tpwd.state.tx.us/landwater/water/environconcerns/hab/ga/workshop/intro.phtml

Graham, J. L., Jones, J. R., Jones, S. B., Downing, J. A. & Clevenger, T. E. (2004). Environmental Factors Influencing Microcystin Distribution and Concentration in the Midwestern United States. *Water Research, 38,* 4395-404.

GWRC. (2004). Management Strategies for Algal Toxins. *Report of the GWRC Research Strategy Workshop, 48.*

Haney, J. F. & Ikawa, M. (2000). A survey of 50 NH Lakes for Microcystins (MCs). 60 http://www.wrrc.unh.edu/pubs/reports/rr65.pdf

HARR-HD. (2006). Harmful Algal Research and Response: A Human Dimensions Strategy. National Office for Marine Biotoxins and Harmful Algal Blooms. M. Bauer, (Ed.) Woods Hole Oceanographic Institution, Woods Hole, MA., 58
http://coastalscience.noaa.gov/stressors/extremeevents/hab/HDstrategy.pdf

HARRNESS. (2005). Harmful Algal Research and Response: A National Environmental Science Strategy 2005-2015. J. S., Ramsdell, D. M.

Anderson, & P. M. Glibert, (Eds.) *Ecological Society of America*, Washington, D.C., 96.

Havens, K. E. (2008). Cyanobacteria blooms: effects on aquatic ecosystems. In: H. K. Hudnell, (Ed.) Cyanobacterial Harmful Algal Blooms: State of the Science and Research Needs. *Advances in Experimental Medicine & Biology, Vol. 619*, Springer. 500.

Hilborn, E. D., Fournie, J. W., Azevedo, S. M., Chernoff, N., Falconer, I. R., Hooth, M. J., Jensen, K., MacPhail, R. & Stewart, I. (2008). Human health effects workgroup report. In: H. K. Hudnell, (Ed.) Cyanobacterial Harmful Algal Blooms: State of the Science and Research Needs. *Advances in Experimental Medicine & Biology, Vol. 619*. Springer. 500.

Hindman, S. H., Favero, M. S., Carson, L. A., Peterson, N. J., Schonberger, L. B. & Solano, J. T. (1975). Pyrogenic reactions during haemodialysis caused by extramural endotoxin. *Lancet, 2*, 732-734.

H. K. Hudnell, (Ed.). (2008). *Cyanobacterial Harmful Algal Blooms: State of the Science and Research Needs. Advances in Experimental Medicine & Biology. Vol. 619*. Springer. 500.

J., Huisman, H. C. P. Matthijs, & P. M.Visser, (eds.). (2005). Harmful Cyanobacteria. *Aquatic Ecology Series. Vol. 3*. Springer, Berlin Heidelberg New York. 243.

Humpage, A. (2008). Toxin types, toxicokinetics, and toxicodynamics. In: H. K. Hudnell, (Ed.) Cyanobacterial Harmful Algal Blooms: State of the Science and Research Needs. *Advances in Experimental Medicine & Biology. Vol. 619*, Springer. 500.

IARC. (2006). Carcinogenicity of nitrate, nitrite, and cyanobacterial peptide toxins. *The Lancet Oncology, 7*, 628-629.

Ibelings, B. W., Havens, K., Codd, G. A., Dyble, J., Landsberg, J., Coveny, M., Fournie, J. W. & Hilborn, E. D. (2008). In: H. K. Hudnell, (Ed.) Cyanobacterial Harmful Algal Blooms: State of the Science and Research Needs. *Advances in Experimental Medicine & Biology, Vol. 619*, Springer. 500.

Izaguirre, G. (2008). Harmful algal blooms and cyanotoxins in Metropolitan Water District's reservoirs. In: H. K. Hudnell, (Ed.) Cyanobacterial Harmful Algal Blooms: State of the Science and Research Needs. *Advances in Experimental Medicine & Biology, Vol. 619*, Springer. 500.

Jewett, E. B., Lopez, C. B., Dortch, Q. Etheridge, & S. M. (2007). National Assessment of Efforts to Predict and Respond to Harmful Algal Blooms in U.S. Waters. Interim Report. Interagency Working Group on Harmful

Algal Blooms, Hypoxia, *and Human Health of the Joint Subcommittee on Ocean Science and Technology*, Washington, DC.

Jochimsen, E. M., Carmichael, W. W., An, J .S., Cardo, D. M., Cookson, S. T., Holmes, C. E. M., Antunes, M. B., de Melo Filho, D. A., Lyra, T. M., Barreto, V. S., Azevedo, S. M. & Jarvis, W. R. (1998). Liver failure and death after exposure to microcystins at a hemodialysis center in Brazil. *New England Journal of Medicine, 338(13)*, 873-878.

Koeing, R. (2006). The pink death: Die-offs of the lesser flamingo raise concern. *Science, 313*, 1734-1725.

Lawrence, J. F., Neidzwiadeck, B., Menard, C., Lau, B. P. Y., Lewis, D., Kuper-Goodman, T., Carbone, S. & Holmes, C. (2001). Comparison of liquid chromatography/ mass spectrometry, ELISA, and phosphatase assay for the determination of microcystins in blue-green algae products. *J AOAC Int., 84*, 1035-1044.

Lowe, R. L. & Pillsbury, R. W. (1995). Shifts in benthic algal community structure and function following the appearance of zebra mussels (Dreissena polymorpha) in Saginaw Bay, Lake Huron. *J. Great Lakes Res., 21*, 558-566.

Miller, A. P. & Tisdale, E. S. (1931). Epidemic of intestinal disorders in Charleston, West Virginia, occurring simultaneously with unprecedented water supply conditions. *American Journal of Public Health, 21*, 198-200.

Murch, S. J., Cox, P. A. & Banack, S. A. (2004). A mechanism for slow release of biomagnified cyanobacterial neurotoxins and neurodegenerative disease in Guam. *Proceedings of the National Academy of Sciences, 101 (33)*, 12228-12231.

Neilan, B. A., Pearson, L. A., Moffitt, M. C., Mihali, K. T., Kaebernick, M., Kellmann, R. & Pomati, F. (2008). The genetics and genomics of cyanobacterial toxicity. In: H. K. Hudnell, (Ed.) Cyanobacterial Harmful Algal Blooms: State of the Science and Research Needs. *Advances in Experimental Medicine & Biology. Vol. 619*, Springer. 500.

NHMRC. (2004). *National Water Quality Management Strategy*, Australian Drinking Water Guidelines.

NHNRC. (2006). *Guideline for Managing Risks in Recreational Waters*. Australian Government.

Nolen, S. L., Carroll, J. H., Combs, D. L., Staves, J. C. & Veenstra, J. N. (1989). Limnology of Tenkiller Ferry Lake, Oklahoma, 1985-1986. In: *Proceedings of the Oklahoma Academy of Science, Vol. 69*, 45-55.

NRA. (1990). *Toxic blue-green algae*. Water Quality Series 2. National Rivers Authority, London, United Kingdom.

Oh, C. O. & Ditton, R. B. (2005). Estimating the economic impacts of golden alga (*Prymnesium parvum*) *on recreational fishing at Possum Kingdom Lake*, Texas. Texas PWD RP T3200-1168. 31.

Oshima, Y., Minami, H., Takano, Y. & Yasumoto, T. (1989). Ichthyotoxins in a freshwater dinoflagellate Peridinium polonicum. In: T., Okaichi, D. M. Anderson, & T. Nemoto, (Eds.) Red tides: *Biology, environmental science and toxicology*. New York, NY, Elsevier Science Publishing Co., 375-377.

Paerl, H. W. (2008). Nutrient and other environmental controls of harmful cyanobacterial blooms along the freshwater-marine continuum. In: H. K. Hudnell, (Ed.) Cyanobacterial Harmful Algal Blooms: State of the Science and Research Needs. *Advances in Experimental Medicine & Biology, Vol. 619*, Springer. 500.

Paerl, H. W., Fulton, R. S., Mosiander, P. H. & Dyble, J. (2001). Harmful Freshwater Algal Blooms with an emphasis on Cyanobacteria. *The Scientific World, 1*, 76-113.

Paerl, H. W. & Fulton, R. S. (2006). Ecology of Harmful Cyanobacteria. In: E. Granéli, & J. T. Turner, (Eds.) Ecology of Harmful Algae. *Ecological Studies, Vol. 180*, 95-109.

Peñaloza, R., Rojas, M., Vila, I. & Zambrano, F. (1990). Toxicity of a soluble peptide from *Microcystis* sp. *to zooplankton and fish. Freshwater Biology. 24*, 233-240.

Perovich, G., Dortch, Q. & Goodrich, J. (2008). Causes, Prevention, and Mitigation Work Group Report. In: H. K. Hudnell, (Ed.) Cyanobacterial Harmful Algal Blooms: State of the Science and Research Needs. *Advances in Experimental Medicine & Biology, Vol. 619*, Springer. 500.

Phlips, E. J., Bledsoe, E., Cichra, M., Badylak, S. & Frost, J. (2002). The distribution of potentially toxic cyanobacteria in Florida. In: D. Johnson, & R. D. Harbison, (Eds.) *Proceedings of Health Effects of Exposure to Cyanobacteria Toxins: State of the Science August 13-14, 2002*, 22-29. http://www.doh.state.fl.us/Environment/community/aquatic/pdfs/Cyanobacteria_200208.pdf

Pilotto, L. S. Douglas, R. M., Burch, M. D., Cameron, S., Beers, M., Rouch, G. R., Robinson, P., Kirk, M., Cowie, C. T., Hardiman, S., Moore, C. & Attewell, R. G. (1997). Health effects of recreational exposure to cyanobacteria (blue-green algae) during recreational water-related activities. Aust. N. Zealand *J. Public Health, 21*, 562-566.

Prepas, E. E., Kotak, B. G., Campbell, L. M. Evans, J. C., Hrudey, S. E. & Holmes, C. F. B. (1997). Accumulation and elimination of cyanobacterial hepatotoxins by the freshwater clam *Anodonta grandis simpsoniana*. *Canadian Journal of Fisheries and Aquatic Sciences*, *54*, 41-46.

Reifel, K. M., McCoy, M. P., Rocke, T. E., Tiffany, M. A., Hurlbert, S. H. & Faulkner, D. J. (2002). Possible importance of algal toxins in the Salton Sea, California. *Hydrobiologia*, *473*, 275-292.

Rinta-Kanto, J. & Wilhelm, S. W. (2006). Diversity of microcystin-producing cyanobacteria in spatially isolated regions of Lake Erie. Applied *Environmental Microbiology*, *72(7)*, 5083-5085.

Rodger, H. D., Turnbull, T., Edwards, C. & Codd, G. A. (1994). Cyanobacterial (blue-green algal) bloom associated pathology in brown trout, Salmo-trutta L, in Loch Leven, Scotland. *Journal of Fish Diseases*, *17*, 177-181.

Roelke, D. L., Errera, R. M., Kiesling, R., Brooks, B. W., Grover, J. P., Schwierzke, L., Urena-Montoya, F., Baker, J. & Pinckney, J. L. (2007). Effects of nutrient enrichment, barley straw extract, and *Prymnesium parvum* immigration on phytoplankton dynamics and toxicity to fish. *Aquatic Microbial Ecology*, *46*, 125-140.

Roset, J., Gibello, A., Aguayo, S., Domínguez, L., Álvarez, M., Fernández-Garayzabal, Zapata, A. & Muñoz, M. J. (2002). Mortality of rainbow trout [*Oncorynchus mykiss* (Walbaum)] associated with freshwater dinoflagellate bloom [*Peridinium polonicum* (Woloszynska)] in a fish farm. *Aquaculture Research*, *33*, 141-145.

Sandifer, P., Sotka, C., Garrison, D. & Fay, V. (2007). Interagency Oceans and Human Health Research Implementation Plan: *A Prescription for the Future. Interagency Working Group on Harmful Algal Blooms, Hypoxia, and Human Health of the Joint Subcommittee on Ocean Science and Technology*, Washington, DC.

Schwimmer, M. & Schwimmer, D. (1968). Medical aspects of phycology. In: D. F. Jackson, (Ed.) *Algae, man and the environment*. Syracuse, NY, Syracuse University Press, 279-358.

Sengco, M. R., Hagstrom, J. A., Graneli, E. & Anderson, D. M. (2005). Removal of *Prymnesium parvum* (Haptophyceae) and its toxins using clay minerals. *Harmful Algae*, *4*, 261-274.

Spaulding, S. & Elwell, L. (2007). Increase in nuisance blooms and geographic expansion of the freshwater diatom Didymosphenia geminata: Recommendations for response. White Paper. http:// www.epa.gov/region8/ water/didymosphenia/White%20Paper%20Jan%202007.pdf

SSERG. (2001). Reconnaissance of the biological limnology of the Salton Sea: final report. Salton Sea Ecosystem Research Group, Department of Biology and Center for Inland Waters, San Diego State University, *San Diego*, California. 1100.

Steffensen, D. A. (2008). Economic cost of cyanobacterial blooms. In: H. K. Hudnell, (Ed.) Cyanobacterial Harmful Algal Blooms: State of the Science and Research Needs. *Advances in Experimental Medicine & Biology. Vol. 619*, Springer. 500.

Stewart, I. (2008). Cyanobacterial poisoning in livestock, wild animals, and birds – an overview. In: H. K. Hudnell, (Ed.) Cyanobacterial Harmful Algal Blooms: State of the Science and Research Needs. *Advances in Experimental Medicine & Biology. Vol. 619*, Springer. 500.

Tango, P., Butler, W. & Michael, B. (2008). Cyanotoxins in the tidewaters of Maryland's Chesapeake Bay: The Maryland Experience. In: Hudnell, H.K. (ed.) Cyanobacterial Harmful Algal Blooms: State of the Science and Research Needs. *Advances in Experimental Medicine & Biology. Vol. 619*, Springer. 500.

Teixera, M. G. L. C., Costa, M. C. N., Carvalho, V. L. P., Pereira, M. S. & Hage, E. (1993). Gastroenteritis epidemic in the area of the Itaparica Dam, Bahia, Brazil. *Bulletin of the Pan American Health Organization*, 27, 244-253.

Tencalla, F. G., Dietrich, D. R. & Schlatter, C. (1994). Toxicity of *Microcystis aeruginosa* peptide toxin to yearling rainbow trout (Oncorhynchus mykiss). *Aquatic Toxicology, 30(3)*, 215-224.

Tucker, C. S. (2000). Off-flavor problems in aquaculture. *Reviews in Fisheries Science, 8(1)*, 45-88.

Turner, P. C., Gammie, A. J., Hollinrake, K. & Codd, G. A. (1990). Pneumonia associated with cyanobacteria. *British Medical Journal, 300*, 1440-1441.

Turner, R. E., Dortch, Q. & Rabalais, N. N. (2004). Inorganic nitrogen transformations at high loading rates in an oligohaline estuary. *Biogeochemistry, 68*, 411-422.

Walker, S. R., Lund, J. C., Schumacher, D. G., Brakhage, P. A., McManus, B. C., Miller, J. D., Augustine, M. M., Carney, J. J., Holland, R. S., Hoagland, K. D., Holz, J. C., Barrow, T. M., Rundquist, D. C. & Gitelson, A. A. (2008). Nebraska Experience. In: H. K. Hudnell, (Ed.) Cyanobacterial Harmful Algal Blooms: State of the Science and Research Needs. *Advances in Experimental Medicine & Biology, Vol. 619*, Springer. 500.

Watson, S. (2001). Literature review of the microalga *Prymnesium parvum* and its associated toxicity. Texas PWD RP T3200-1158. 38.

Wehr, J. D. & R. G. Sheath, (Eds.) (2003). Freshwater Algae of North America. Academic Press. 918.

Westrick, J. A. (2008). Cyanobacterial toxin removal in drinking water treatment processes and recreational waters. In: H. K. Hudnell, (Ed.) Cyanobacterial Harmful Algal Blooms: State of the Science and Research Needs. *Advances in Experimental Medicine & Biology, Vol. 619*, Springer. 500.

Williams, C. D., Burns, J., Chapman, A., Flewelling, L, Pawlowicz, M. & Carmichael, W. W. (2001). *Assessment of cyanotoxins in Florida's lakes*, reservoirs, and rivers. Palatka, FL: St. Johns River Water Management District.

WHO. (1998). Guidelines for drinking water quality. Second edition, Addendum to Volume 2, Health criteria and other supporting information. World Health Organization. Geneva, Switzerland.

WHO. (1999). Toxic cyanobacteria in water: A guide to their public health consequences, monitoring and management. I. Chorus, & J. Bartram, (Eds.) World Health Organization, E & FN Spon, New York, 416.

WHO. (2000). Monitoring Bathing Waters. *A Practical Guide to the Design and Implementation of Assessments and Monitoring Programmes.*

WHO. (2003). *Guidelines for safe recreational water environments. Volume 1, Coastal and fresh waters.* World Health Organization. Geneva, Switzerland. *219.*

Yu, S .Z. (1995). Primary prevention of hepatocellular carcinoma. *Journal of Gastroenterology and Hepatology, 10*, 674-682.

Zimba, P. V., Khoo, L., Gaunt, P., Carmichael, W. W. & Brittain, S. (2001). Confirmation of Catfish Mortality from *Microcystis* toxins. *Journal of Fish Diseases, 24*, 41-47.

Zimba, P. V., Rowan, M. & Triemer, R. (2004). Identification of euglenoid algae that produce Ichthyotoxins. *Journal of Fish Diseases, 27*, 115-117.

In: Marine and Freshwater Harmful Algal Blooms   ISBN: 978-1-60741-838-2
Editor: Peter E. Williams   © 2010 Nova Science Publishers, Inc.

Appendix I

FEDERAL FRESHWATER/ INLAND HAB RESEARCH AND RESPONSE PROGRAMS

Joint Subcommittee on Ocean Science and Technology

Agricultural Research Service

- Studying effects of land management on nutrient losses from crop land to surface water and developing best management practices that decrease water export of nitrogen and phosphorus. (Extramural)
- Developing control methods to selectively eliminate cyanobacteria from phytoplankton communities inhabiting ponds used for catfish aquaculture and from California rice fields. (Extramural)
- Researching use of wetlands to treat swine wastewater. (Extramural)

U.S. DEPARTMENT OF COMMERCE

National Oceanographic and Atmospheric Administration

Center for Coastal Fisheries and Habitat Research (CCFHR)

- Sampling and mapping cells and toxin levels in Great Lakes in collaboration with GLERL.

- Performing genetic analysis of *Microcystis* strains in the Great Lakes to determine the occurrence of toxigenic strains.
- (*future plans*) Continue monitoring of *Microcystis* toxin production in the Great Lakes, which can serve as a source of data for the exposure assessment component of epidemiological studies.
- (*future plans*) Develop enhanced molecular tools for assessing the ability of *Microcystis* blooms to produce toxins.
- (*future plans*) Develop accurate satellite monitoring capabilities to monitor and track blooms.

Center for Coastal Monitoring and Assessment (CCMA)

- Providing real-time products from the Sea-viewing Wide Field-of-view Sensor (SeaWiFS) satellite since 2004, which are guiding sampling efforts for *Microcystis* in western Lake Erie and Saginaw Bay.
- Developing SeaWiFS and Medium-spectral resolution imaging spectrometer (MERIS) imagery to aid in detection of cyanobacterial blooms.
- Considering the development of a HAB forecasting system for the Great Lakes with GLERL.

Center for Sponsored Coastal Ocean Research (CSCOR)

- Ecology of Harmful Algal Blooms (ECOHAB) (Extramural)
  - Determining the role of nutrients and zebra mussels in promoting *Microcystis* blooms, in collaboration with EPA.
  - Researching the fundamental ecology of *Microcystis*.
  - Measuring bioaccumulation and impacts of cyanotoxins in food web.
- Monitoring and Event Response for Harmful Algal Blooms (MERHAB) (Extramural)
  - Applying new tools to detect, track, map, predict, and respond to blooms of cyanobacteria and their associated toxins through MERHAB Lower Great Lakes project, a comparative regional study.

- Established a toxin analysis facility at SUNY-ESF, which is one of the few facilities in the United States that can cover all known cyanobacterial toxins.

Great Lakes Environmental Research Laboratory (GLERL), Center for Excellence in Human Health

- Exploring environmental and genetic factors affecting *Microcystis* cells and toxin production in Lakes Erie and Huron. (Extramural)
- (*future plans*) Research the genomics of microcystin synthetase genes. (Extramural)
- (*future plans*) Study the sublethal impacts of microcystin accumulation in fish. (Extramural)
- Mapping spatial and temporal variability in microcystin concentrations in western Lake Erie and Saginaw Bay in collaboration with CCFHR.
- Maintaining HAB Event Response website and listserver to inform other researchers and the public about the presence of HAB taxa and current toxin concentrations.
- Identifying environmental factors that promote CyanoHABs.
- Developing genetic tools for detecting the presence of toxin-producing HAB strains.
- Conducting epidemiological study on recreational exposure to HABs in Michigan in collaboration with CDC and Mote Marine Laboratory.
- Developing the capacity for forecasting the presence of toxic HABs.
- Measuring accumulation of microcystins in fish tissue.
- Evaluating the role of zebra mussels in promoting *Microcystis* blooms.

Oceans and Human Health Initiative (OHHI)

- Exploring emerging HAB threats in the Great Lakes: collecting and characterizing novel toxigenic cyanobacteria from Great lakes and other freshwater systems, establishing molecular and biochemical diagnostics for their presence and presence of their toxins, and

- generating a series of toxin and cell standards to be used by GLERL. (Extramural)
- Determining the distribution and occurrence of anatoxin-a and cylindrospermopsin-producing organisms and their toxins in Lake Erie. (Extramural)
- Evaluating use of high-throughput assays for detection of cyanotoxins. (Extramural)
- Creating training and reference materials for GLERL. (Extramural)

Sea Grant

- Researching the link between algal mats and fecal bacteria and how to remove cyanotoxins from drinking water. (Extramural)

U.S. DEPARTMENT OF DEFENSE

United States Army Medical Research Institute for Infectious Diseases (USAMRIID)

- Developing new electrochemiluminescence-based immunoassay for microcystins.
- Using High Performance Liquid Chromatography method to test for saxitoxins.

Army Corps of Engineers (USACE)

- Managing approximately 400 reservoirs, several of which are impacted by HABs.
- Conducting research to support management of these reservoirs.
  - Evaluating and developing new chemicals for freshwater HAB control.
  - Developing a framework for managing HABs.
  - Assessing risks of algal toxins in freshwater systems.
- Collaborating with EPA – NCEA and Aquatic Ecosystem Restoration Foundation.

U.S. DEPARTMENT OF INTERIOR

Bureau of Reclamation (BOR)

- Collaborating with USFWS to research the potential impact of algal blooms on aquatic species in Upper Klamath Lake, California.

U.S. Fish and Wildlife Service (USFWS)

- Documenting toxic CyanoHAB events involving birds, including collection of carcasses and water samples for toxin analysis.

- **New Mexico Ecological Services Environmental Contaminants Program**
  - Tracking spread of golden alga from Texas to New Mexico.
  - Funding local initiatives in Texas to study *P. parvum*. (Extramural)

U.S. Geological Survey (USGS)

- **Columbia Environmental Research Center**
  - Measuring microcystin concentrations in reservoirs and wetlands and linking results to water quality and fish toxicity.
  - Developing sample collection and analytical techniques for measuring cyanotoxins with Kansas Water Science Center.

- **Kansas Water Science Center**
  - Developing tools to predict onset of cyanobacterial-related taste-and-odor episodes in drinking water reservoirs.
  - Developing sample collection and analytical techniques for measuring cyanotoxins with Columbia Environmental Research Center.

- **Lake Michigan Ecological Research Station and the Great Lakes Science Center**
  - Researching relationship between algal mats and fecal indicator bacteria.

- **National Wildlife Health Center**
  - Analyzing samples from birds, mammals, and reptiles to document effects of biotoxins; maintaining extensive database of wildlife mortality investigations.

- **Western Fisheries Research Center, Klamath field station**
  - Studying, in partnership with the BOR, impact of algal blooms on aquatic species in Upper Klamath Lake to protect endangered fish.

- **National Water Quality Assessment Program (NAWQA)**
  - Performing ecological studies in 42 watersheds as part of monitoring long term trends to see what factors affect water quality.

DEPARTMENT OF HEALTH AND HUMAN SERVICES

Centers for Disease Control and Prevention (CDC)

- Developing public health responses to freshwater HABs through cooperative agreements with five coastal states (Florida, Maryland, North Carolina, South Carolina, and Virginia).
- Assessing human exposure/self-reported symptoms associated with recreational activities in lakes with algal blooms.
- Developing HABISS, a web-based system for state health departments to collect both medical information about human and animal illnesses and physical/biological information about bloom events.
- (*future plans*) Conduct additional studies to assess human exposures to HAB toxins during recreational activities, such as swimming in lakes with HABs.

- (*future plans*) Participate in a multi-agency, international (e.g., with Australia) group to study the public health effects of exposure to cyanobacterial toxins.
- (*future plans*) Use data from the Great Lakes Observing System component of the ocean observing systems to assess how waterborne risks affect lakeshore communities.

U.S. Food and Drug Administration (FDA)

- Ensuring the safety of bottled water, dietary supplements, and seafood from aquaculture.
- Providing reference standard saxitoxin.
- Conducting methods development for saxitoxin and, to a lesser extent, microcystins.

National Institute of Environmental Health Sciences (NIEHS)

- Using the zebrafish embryo model in developmental toxicity assays to understand the emerging role of freshwater cyanobacterial toxins, as well as cyanobacterial lipopolysaccharides, in environmental health using cyanobacterial isolates from the ecologically unique and distinct freshwater microbial communities of the Florida Everglades (and associated waterways in South Florida) and the Northern Great Lakes. (Extramural)

U.S. ENVIRONMENTAL PROTECTION AGENCY

- Regulating safety of U.S. drinking water.
- Acting as liason (via regional offices) between state authorities and Federal responders if a HAB problem occurs.
- Researching nutrient criteria development.
- Creating guidance on methods for controlling and mitigating algal growth within water treatment plants.
- Establishing criteria for laboratories providing algal toxin analyses to drinking water utilities.

- Funded Phase I of a SBIR research project to develop a faster, simpler fiber optic probe for detecting microcystin-LR in the field. (Extramural)
- (*future plans*) Develop new detection methods for cyanobacterial cells and toxins, such as microcystins, cylindrospermopsin, and anatoxin-a, in drinking water.
- (*future plans*) Study health effects of exposure to cyanobacterial toxins found in drinking water, such as microcystins, cylindrospermopsin, and anatoxin-a.

ECOHAB

- Determining role of nutrients and zebra mussels in promoting *Microcystis* blooms, in collaboration with NOAA. (Extramural)

Gulf of Mexico Program

- Funding research to test water, crustaceans, and shellfish consumed by humans for toxins produced by CyanoHABs. (Extramural)

Office of Research and Development (ORD)

- Conducting or supporting research through four national centers or laboratories -- the National Center for Environmental Research (NCER), the National Exposure Research Laboratory (NERL), the National Health and Environmental Effects Research Laboratory (NHEERL), and the National Risk Management Research Laboratory (NRMRL). (Some extramural)

National Center for Environmental Research (NCER)

- Supporting research in the academic community to develop gene microarray assays for monitoring of cyanobacteria and cyanotoxins in drinking water.

- Producing a microarray suitable for use as a tool of risk assessment of cyanobacteria and cyanobacterial toxins in drinking water reservoirs and lakes.

National Exposure Research Laboratory (NERL)

- Exploring the use of titanium dioxide, an emerging "green" technology, for the treatment of microcystins in drinking water.
- Developing techniques for separation, detection, identification and quantitative measurement of six cyanobacterial toxins including anatoxin-a, cylindospermopsin, and four microcystins (microcystin-RR, -LR, -YR, -LA).
- Conducting risk and impact assessment for freshwater HABs through the National Center for Environmental Assessment (NCEA), a critical link between researchers within ORD and EPA decision makers.

National Center for Environmental Assessment (NCEA)

- Conducting risk and impact assessment for microcystin exposure.
- Prepared draft *Toxicological Reviews of Cyanobacterial Toxins: Anatoxin-a, Cylindrospermopsin and Microcystins LR, RR, YR and LA*, which are a series of dose-response studies to support the health assessment of unregulated contaminants on the CCL and to aid the EPA/ Office of Water in regulatory decision making. The reviews were published in the Federal Register and were available for public comment to be followed by scientific peer review: http://cfpub.epa.gov/ncea/cfm/recordisplay.cfm?deid=160546, http://cfpub.epa.gov/ncea/cfm/recordisplay.cfm?deid=160547, http://cfpub.epa.gov/ncea/cfm/recordisplay.cfm?deid=160548.

NATIONAL SCIENCE FOUNDATION

- Supporting research on the following general topics related to CyanoHABs (Extramural):
  - analyzing the genetics of nitrogen fixation,
  - developing tools to elucidate metabolic and regulatory networks in cyanobacteria,

- investigating the vertical distribution of phytoplankton in lakes using mathematical models,
- identifying the role of dissolved organic material in regulating primary production in lakes,
- developing a simple model in Lake Erie to examine system dynamics and role of human-biophysical coupling to understand better "surprise events" such as HABs, dead zones, or game fish kills,
- researching reactive observing systems, which can monitor concentrations of microorganisms and algae in water and reconfigure sampling depending on measurements taken, and
- studying native versus invasive cyanobacteria species and how nitrogen fixation benefits the invasive species.

• Supports the UTEX Culture Collection of Algae at the University of Texas at Austin which contains approximately 3,000 different strains of living algae, representing most major algal taxa including some freshwater species. (Extramural)

In: Marine and Freshwater Harmful Algal Blooms   ISBN: 978-1-60741-838-2
Editor: Peter E. Williams   © 2010 Nova Science Publishers, Inc.

Appendix II

OTHER NATIONAL PROGRAMS

Joint Subcommittee on Ocean Science and Technology

AMERICAN WATER WORKS ASSOCIATION RESEARCH FOUNDATION (AWWARF)

AwwaRF is one of the primary bodies in the United States interested in identifying and funding drinking water-related cyanotoxins research and development. The Foundation is comprised of, and largely funded by, member organizations that voluntarily subscribe in order to support and benefit from the water-related research that the Foundation sponsors. Close to 900 water utilities and more than 50 water-related consulting firms and manufacturing companies currently subscribe to the Foundation. The majority of subscribers are in the United States, but others are located in Canada, Australia, and Europe. Since its inception, the Foundation has sponsored more than 600 completed research projects. The Foundation's sponsored research is largely funded by subscribers, but some are also funded by the U.S. government. http://www.awwarf.org/theFoundation/

NATIONAL HAB COMMITTEE (NHC)

NHC has been established as a critical component for implementation of the HARRNESS (2005) plan. The NHC is an elected body with members who represent the HAB research and management community at the national level. The NHC serves as an important link between Federal programs and organizations involved in HAB research and management.

THE U.S. NATIONAL OFFICE FOR HARMFUL ALGAL BLOOMS

The U.S. National Office for HABs (http://www.whoi.edu/redtide/), funded by NOAA's CSCOR, provides critical coordination and technical support capabilities that enhance the Nation's ability to respond to and manage the growing threat posed by HABs. It also serves as an important liaison with the scientific community and related national and international programs. The National Office maintains a website (i.e., the "Harmful Algae Page") for distribution of important information on HABs, including freshwater HABs.

In: Marine and Freshwater Harmful Algal Blooms   ISBN: 978-1-60741-838-2
Editor: Peter E. Williams   © 2010 Nova Science Publishers, Inc.

Appendix III

STATE AND LOCAL INITIATIVES

Joint Subcommittee on Ocean Science and Technology

ARIZONA

- City of Phoenix Water Services Department and Arizona State University are developing a monitoring and response program to assess toxin and taste-and-odor problems in distribution systems.

CALIFORNIA

- Task force of county, state, Federal and tribal authorities convened to develop guidance.
- Department of Health Services provides extensive web information including geographic location of past and current blooms (http://www.dhs.ca.gov/ps/ddwem/bluegreenalgae/).
- Metropolitan Water District of Southern California is investigating drinking water issues. Research includes predictive tools for early warning and management of taste-and-odor events.

FLORIDA

- Department of Health (DOH) Aquatic Toxins Program provides fact sheets and has the capacity to respond to CyanoHAB events that pose potential human health threats (http://www.doh.state.fl.us/environment/ community/aquatic/cyanobacteria.htm).
- DOH maintains a hotline through the Poison Control Center in Miami for reporting illnesses due to marine or freshwater toxin exposure (http://www.doh.state.fl.us/environment/community/aquatic/index.html).
- South Florida Water Management District conducts water sampling in Florida lakes.
- Cities of Cocoa and Melbourne and the St. John's Water Management District are characterizing treatment technologies, such as ozone, oxidation, adsorption, and PAC and membrane filtration, for removal of algal toxins.

INDIANA

- Department of Environmental Management, along with Soil and Water Conservation Districts, conducts sampling for CyanoHABs (http://www.in.gov/dnr/fishwild/fish/cylind.htm).

IOWA

- Department of Natural Resources and Iowa State University monitor 132 lakes for cyanobacteria and associated toxins.

MARYLAND

- Department of Natural Resources provides extensive information on the web, including forms for reporting new blooms (*http://www.dnr.state.md.us/bay/hab/index.html*).

- Department of Natural Resources and Chesapeake Bay Program issue a *Microcystis* bloom forecast for the Potomac River.

MICHIGAN

- Department of Environmental Quality provides web information (http://www.deq.state.mi.us/documents/deq-ead-tas-algae.pdf).

MINNESOTA

- Pollution Control Agency provides web information (*http://www.pca.state.mn.us/water/clmp-toxicalgae.html*).

NEBRASKA

- Natural Resource Districts, Public Power District, Game and Parks Commission, and Department of Environmental Quality, in collaboration with USACE, monitor CyanoHABs in lakes and inform the public via extensive web information (*http://www.ndeq.state.ne.us/*).

NEW HAMPSHIRE

- Department of Environmental Services provides web information about cyanotoxin effects via recreational exposure and a hotline for reporting suspected blooms (http://www.des.state.nh.us/Beaches/cyanobacteria.html).

NEW YORK

- Department of Public Health, Bureau of Toxic Substance Assessment maintains web information (http:// www. health. state.ny .us/ nysdoh/ water/ bluegreenalgae.htm).
- Department of Environmental Conservation is a partner in the Lake Champlain Basin Program, which issues alerts on CyanoHABs in the lake and posts a map of the lake with existing bloom status.

NEVADA

- Department of Environmental Protection led an Algae Task Force which worked to understand the causal factors and provided early detection, education and outreach, and prevention after a 2001 bloom in Lake Mead.

NORTH CAROLINA

- Department of Health and Human Services has an active HAB Program which primarily focuses on education and outreach pertaining to *Pfiesteria* blooms. The program maintains a hotline for reporting suspected blooms and for attaining public health information (http:// www.epi.state.nc.us/epi/hab/bluegreen.html).

OREGON

- Task force of governmental, academic, and private interests convened to develop recommendations for recreational guidance when blooms occur.
- Department of Human Services, Environmental Toxicology Program posts cyanobacteria advisories for lakes when blooms occur and other general public health information (http://www.oregon.gov/DHS/ph/ envtox/ maadvisories.shtml#Lemolo).

TEXAS

- Parks and Wildlife Department conducts statewide surveys to monitor for *P. parvum* and posts information about the location of blooms on the internet (http://www.tpwd.state.tx.us/landwater/ water/environ concerns/hab/).
- Parks and Wildlife Department has also been involved in testing agents to control *P. parvum* in aquaculture facilities.

VERMONT

- Department of Public Health is an active partner in the Lake Champlain Basin Program (see listing for New York above) (http://healthvermont. gov/enviro/bg_algae/bgalgae.aspx).
- Department of Environmental Conservation, in collaboration with academic researchers, conducts water sampling in Lake Champlain.

WASHINGTON

- King County Natural Resources and Parks Department routinely monitors three major lakes for water quality (http://dnr.metrokc.gov/wlr/ waterres/lakes/bloom.htm).

WISCONSIN

- Department of Natural Resources maintains general CyanoHAB information on the internet (http://www.dnr.state.wi.us/org/ land/parks/safety/bluegreenalgae.html).
- Division of Public health provides a fact sheet on cyanobacteria, their toxins, and health impacts (http://dhfs.wisconsin.gov/eh/water/ fs/cyanobacteria.pdf).

Appendix IV

INTERNATIONAL EFFORTS

Joint Subcommittee on Ocean Science and Technology

AOAC MARINE AND FRESHWATER TOXINS TASK FORCE

A Marine and Freshwater Toxins Task Force was developed under the auspices of AOAC International to work on prioritizing, funding, and accelerating validation studies of methods for marine and freshwater toxins. The Task Force addresses the need for validated methods by focusing efforts, setting priorities, and identifying economic and intellectual resources. The Task Force also has a training initiative to assist in method implementation (http://www.aoac.org/marine_toxins/task_force.htm).

CYANONET

A global effort under UNESCO that involves at least 69 countries. This effort collates practical experience of cyanotoxin incidents and available management strategies with a goal to raise awareness and increase the capacity for management in developing countries (http://www.cyanonet.org/).

GLOBAL WATER RESEARCH COALITION (GWRC)

The GWRC is dedicated to promoting international cooperation and collaboration in water-related research. The GWRC was officially formed in April 2002 with the signing of the partnership agreement at the International Water Association's 3rd World Water Congress. The AwwaRF is the U.S. member of the GWRC (http://www.globalwaterresearchcoalition.net/).

INTERNATIONAL CONFERENCE ON TOXIC CYANOBACTERIA

A triennial conference attended by about 200 researchers worldwide to maintain awareness of cyanotoxins research and advancements.

NATIONAL RESEARCH COUNCIL CANADA

The Certified Reference Materials Program at the National Research Council Canada's Institute for Marine Biosciences provides certified standards for some cyanotoxins (http://imb-ibm.nrc-cnrc.gc.ca/crmp/freshwater/cyanobacteria_e.php).

TOXIC-EU

TOXIC-EU is a research project supported by the European Commission under the Fifth Framework Programme to provide problem-driven solutions for water management and purification strategies to reduce the human health risk of cyanobacteria and cyanotoxins in drinking water sources in Europe (http://www.cyanotoxic.com/overview.htm).

WORLD HEALTH ORGANIZATION (WHO)

'Water and health' is an important focus of the WHO, and a major WHO activity is developing guidelines to reduce health risk from exposure to toxic substances in water, including cyanotoxins (http://www.who.int/en/).

INDEX

A

acid, 18, 85
activated carbon, 36, 38
acute, 46, 62
adducts, 62
adsorption, 106
agricultural, 35
aid, 2, 48, 71, 94, 101
Alabama, 25
alanine, 18, 85
algae, ix, x, 7, 19, 25, 27, 30, 33, 39, 40, 41, 42, 43, 47, 48, 54, 67, 71, 84, 88, 89, 92, 102, 107, 109
Algal, i, v, xi, xii, xv, xvii, xviii, 1, 2, 3, 4, 5, 36, 39, 40, 42, 45, 46, 50, 70, 73, 76, 79, 83, 84, 85, 86, 87, 88, 89, 90, 91, 92, 94, 104
amino acid, 18, 85
ammonium, 41
analytical techniques, 44, 97
analytical tools, 44
animal health, 9, 45, 66
antibody, 39
application, 32, 38, 78
aquaculture, 8, 14, 17, 23, 35, 40, 41, 91, 93, 99, 109
aquatic systems, 33
Arizona State University, 38, 105
Arkansas, 25

Army Corps of Engineers, xviii, 4, 96
assessment, 5, 35, 47, 48, 49, 50, 76, 80, 84, 85, 94, 101
assumptions, 42
Australia, 13, 23, 28, 31, 49, 81, 99, 103
availability, xiv, 11, 34, 44, 50, 56, 61, 70, 71
awareness, 4, 20, 43, 73, 80, 110, 111

B

bacteria, 46, 47, 67, 96, 98
bacterial, 34
beaches, 25, 26
benefits, 35, 73, 102
bioaccumulation, 64, 94
bioactive compounds, ix, x, xii, 60, 64
bioconcentration, 64
bioindicators, 55
biomarkers, 62
biomass, ix, x, 7, 9, 10, 18, 25
biosynthesis, 59
biota, 55, 63, 64
birds, 19, 47, 91, 97, 98
Brazil, 13, 15, 82, 88, 91
broad spectrum, 40
Bureau of Reclamation, xvii, 47, 97
by-products, 69

Index

C

Canada, x, 13, 17, 81, 82, 103, 111
cancer, 13, 47
carbon, 36, 38, 67
carcinogenic, 48
carcinogenicity, 61, 62
carcinoma, 92
catfish, 17, 23, 28, 35, 36, 40, 41, 93
CDC, xvii, xviii, 4, 20, 31, 45, 46, 49, 58, 59, 95, 98
cell, 27, 44, 56, 71, 73, 96
Centers for Disease Control, xvii, 4, 98
chemical oxidation, 38
chemicals, 40, 96
children, 43
China, 13, 48, 81
chlorine, 37
chlorophyll, 24, 65
citizens, 26, 73
classes, 44
classical, 60
classification, 48
clay, 40, 90
Clean Water Act, xvii, 2
climate change, 11
closure, 26
coagulation, 37, 38, 40, 69
collaboration, xv, 32, 42, 48, 72, 93, 94, 95, 100, 107, 109, 111
Colorado, 25
Columbia, 33, 97
communication, 74, 75, 76, 77, 78, 79
communities, xv, 23, 28, 40, 49, 72, 73, 77, 93, 99
community, 16, 32, 38, 43, 73, 76, 77, 82, 88, 89, 100, 104, 106
complex interactions, 34, 37
composition, 32, 69
compounds, ix, x, xii, 7, 8, 11, 17, 23, 25, 36, 40, 60, 64, 69
concentration, 27, 44, 65
conductivity, 41
Congress, 5, 111
consensus, 57
consulting, 103
contaminants, x, 2, 48, 78, 101
contamination, 55, 66
control, xiv, 11, 30, 34, 36, 39, 40, 41, 56, 59, 68, 69, 70, 82, 83, 93, 96, 109
copper, 13, 40, 41
cost-effective, 36, 70
costs, xi, 1, 17, 23, 68
Council on Environmental Quality, 3
CRC, 85
critical infrastructure, 79
crustaceans, 33, 73, 100
culture, 33, 71
Cyanobacteria, xvii, 4, 9, 10, 42, 81, 82, 84, 87, 89, 111
cyanobacterium, 39, 46, 83
cytotoxins, 9

D

data collection, 45, 46
database, 33, 47, 56, 71, 72, 80, 81, 98
dead zones, 102
deaths, 10, 14, 15, 19, 20
decision making, 48, 101
decision-making process, 2
degradation, 37, 60
degrading, 37
Denmark, 31
Department of Agriculture, xix, 3, 4
Department of Commerce, 3, 93
Department of Defense, 96
Department of Health and Human Services, 98, 108
Department of Interior, 3, 97
dermatitis, 12
detection, xii, xiii, xiv, 33, 41, 44, 48, 49, 50, 54, 55, 65, 66, 70, 81, 94, 96, 100, 101, 108
detoxification, xii, 60
developing countries, 110
dialysis, 13, 14
Diamond, 19
diatoms, 25
dietary, 99

dinoflagellates, 25
disinfection, 37, 38
distribution, ix, 23, 31, 32, 34, 38, 60, 89, 96, 102, 104, 105
diversity, 32
DNA, 62
dose-response relationship, 47, 54, 61
draft, 2, 48, 101
drinking water, x, xi, 2, 8, 13, 14, 15, 17, 21, 23, 27, 31, 32, 36, 37, 38, 41, 42, 44, 45, 47, 48, 56, 66, 68, 69, 70, 73, 76, 84, 85, 92, 97, 99, 100, 101, 103, 105, 111
duplication, xv, 71, 72, 76, 79, 81
duration, 27, 62

E

E. coli, 25, 47
early warning, xiii, 19, 35, 65, 105
ecological, ix, xi, xii, xiii, 4, 33, 47, 55, 58, 59, 60, 61, 63, 67, 72, 98
Ecological Society of America, 87
ecology, 34, 81, 85, 94
economic losses, 17, 23, 36
ecosystem, x, xii, 7, 10, 50, 56, 64, 67, 70
ecosystems, ix, x, xiii, 9, 18, 19, 28, 35, 47, 61, 65, 67, 83, 87
effluents, 17
ELISA, xvii, 37, 88
embryo, 47, 99
England, 14
environment, xiv, 2, 8, 40, 57, 60, 64, 67, 69, 90, 106
environmental conditions, 10, 34, 56, 60, 66
environmental factors, 34, 37, 45, 49, 67, 68, 95
environmental impact, xiv, 56, 68
Environmental Protection Agency, x, xvii, 2, 3, 99
enzyme-linked immunosorbent assay, 37
EPA, x, xi, xv, xvii, xviii, 2, 4, 15, 23, 31, 32, 34, 35, 36, 38, 44, 48, 50, 51, 54, 58, 59, 80, 82, 84, 94, 96, 101
estuaries, xv, 7, 27, 32, 50, 51
Europe, 103, 111

European Commission, 111
eutrophication, 16, 35
expertise, 52, 56, 70, 72, 73
exposure, xii, xiii, 13, 14, 18, 19, 43, 44, 45, 46, 47, 48, 49, 55, 56, 58, 60, 61, 62, 63, 65, 66, 73, 88, 89, 94, 95, 99, 100, 101, 106, 107, 112

F

failure, 12, 88
farming, 17
fatalities, 84
FDA, xvii, 4, 37, 40, 44, 58, 59, 99
Federal Register, 101
fiber, 45, 100
filtration, 37, 38, 69, 106
Finland, 31
fish, x, xii, 7, 8, 9, 10, 18, 25, 26, 28, 33, 35, 40, 41, 47, 49, 59, 63, 64, 68, 73, 89, 90, 95, 97, 98, 102, 106
Fish and Wildlife Service, xviii, 31, 97
fishing, 22, 28, 74, 89
floating, 39, 42
flocculation, 37, 38, 41
flooding, 24
flora and fauna, 40
flushing, 11, 35, 40, 69
focusing, 78, 110
food, x, xiii, 11, 19, 40, 61, 63, 64, 70, 78, 94
Food and Drug Administration (FDA), 4, 99
forecasting, 41, 42, 49, 54, 94, 95
Forest Service, 19, 20
fresh water, 92
funding, 21, 32, 38, 46, 47, 50, 51, 79, 103, 110
funds, 33
fungal, 78

G

gastroenteritis, 83
gastrointestinal, 13, 14, 17, 18

gene, 38, 45, 59, 100
generation, xi, 1, 23
genes, 45, 49, 59, 60, 95
genetics, 88, 101
Geneva, 92
genomics, 49, 88, 95
Georgia, 11, 25
Germany, 31
goals, xiii, xiv, xv, 52, 53, 54, 55, 56, 57, 59, 61, 77
government, 5, 19, 72, 74, 75, 76, 103
Great Lakes, xi, xv, xvii, 9, 20, 25, 31, 32, 34, 41, 42, 44, 45, 49, 50, 51, 65, 78, 81, 82, 83, 88, 93, 94, 95, 98, 99
groups, ix, 62, 71, 73, 81
growth, x, 8, 19, 37, 50, 56, 67, 71, 99
Guam, 85, 88
guidance, 15, 19, 23, 36, 52, 75, 99, 105, 108
guidelines, xi, 1, 2, 10, 18, 22, 27, 32, 37, 39, 47, 50, 61, 72, 112
Gulf of Mexico, 100

H

harm, 7, 25
harvesting, 17, 39, 40
hazards, ix, xiii, 7, 61
health, ix, xi, xii, 2, 4, 5, 7, 9, 10, 15, 19, 22, 23, 25, 26, 27, 30, 32, 36, 43, 44, 45, 46, 47, 48, 54, 55, 56, 58, 59, 60, 61, 62, 64, 66, 72, 74, 80, 82, 87, 98, 99, 100, 101, 106, 108, 109, 111, 112
Health and Human Services, 98, 108
health effects, xii, 15, 44, 46, 47, 48, 54, 55, 56, 58, 62, 64, 66, 82, 87, 100
hemodialysis, 88
hemorrhage, 12
hepatitis a, 13
hepatocellular carcinoma, 92
hepatotoxins, 15, 90
herbicides, 36
high risk, 40
Holland, 91
Hong Kong, 82

hospitalization, 13
human, ix, xi, xii, xiii, 2, 4, 5, 7, 9, 10, 14, 15, 18, 19, 23, 25, 26, 27, 32, 34, 44, 45, 46, 49, 54, 55, 56, 58, 59, 60, 61, 64, 66, 67, 68, 72, 82, 98, 102, 106, 111
human exposure, 46, 49, 64, 98
humans, ix, x, xiii, 7, 15, 17, 30, 33, 40, 44, 47, 50, 55, 56, 58, 60, 61, 64, 65, 100
hydrodynamic, 42
hydrologic, 11, 29, 34
hydrological, 41
hydrology, ix, 16, 67, 69
hypotension, 14
hypoxia, 1, 5, 7, 9, 10
Hypoxia, xi, xii, xv, xvii, xviii, 1, 2, 3, 5, 76, 88, 90

I

IARC, 48, 87
id, 80
identification, 33, 44, 47, 56, 57, 70, 71, 72, 73, 101
imagery, 32, 94
images, 65, 84
imaging, 94
immigration, 90
impact assessment, 48, 101
implementation, 41, 56, 65, 75, 77, 104, 110
in situ, xii, 33, 57, 66, 72
in vitro, 63
in vivo, 63
incidence, 13
Indiana, 33, 106
indicators, 10, 25, 47, 58, 62, 67
industry, 17, 23, 40
infection, 34
infrastructure, xiv, 53, 55, 56, 59, 61, 70, 75, 77, 79
ingestion, 17, 18
inhalation, 18, 62
Innovation, xviii, 45
institutions, xi, 50
instruments, 57, 66
integration, 72, 75

Index

interaction, xiv, xv, 67, 75
interactions, 34, 37, 60, 75
International Agency for Research on Cancer, 47
international communication, 74
internet, 43, 109
invasive species, 102
invertebrates, 63
Investigations, 69
investment, xii, 52
IOC, 79, 80
irritation, 14

J

Japan, 10, 28, 81

K

killing, xii, 59
King, 109

L

Lake Pontchartrain, 24, 85
lakes, 8, 9, 19, 20, 22, 27, 28, 33, 34, 36, 39, 43, 45, 49, 84, 92, 95, 98, 101, 102, 106, 107, 108, 109
large-scale, 60
legislation, xi, 50, 76
leukocytes, 62
life cycle, 34
likelihood, 41
lipopolysaccharides, 47, 99
liquid chromatography, 88
liver, 12, 17
livestock, 19, 42, 91
local government, 5, 72, 74
London, 85, 89
losses, 17, 23, 28, 93
Louisiana, 24, 32
Louisiana State University, 24
low-level, xii, 17, 60

M

macroalgae, 25
Maine, 78
maintenance, 9, 34
management, xii, xiv, xv, 4, 6, 9, 11, 16, 31, 33, 34, 35, 41, 47, 50, 52, 57, 67, 68, 69, 70, 72, 73, 74, 77, 78, 80, 81, 83, 84, 92, 93, 96, 104, 105, 110, 111
management practices, 35, 41, 69, 80, 93
manipulation, 40
manufacturing, 103
Maryland, xviii, 8, 26, 33, 43, 74, 91, 98, 106
mass spectrometry, 88
Massachusetts, 78
measurement, 44, 101
membership, 77
merit-based, xv, 50
metabolic, 34, 60, 101
metabolic pathways, 60
metabolism, xii, 55, 58, 59
metabolites, 57, 62
Mexico, 8, 25, 33, 97
Miami, 106
microalgae, 25
microarray, 38, 45, 100, 101
microbial communities, 99
microorganisms, 33, 102
minerals, 90
Minnesota, 107
missions, 53
Mississippi, 24, 85
modeling, 35
models, xiv, 34, 41, 42, 56, 64, 66, 67, 102
modules, 45, 46
molecular biology, 81
mortality, 98
muscle, 14

N

NASA, 58, 59
nation, 36

National Academy of Sciences, 85, 88
National Aeronautics and Space Administration, xv, 50
National Institutes of Health, 4
National Oceanic and Atmospheric Administration (NOAA), 4
National Research Council, 111
National Science Foundation, xv, xviii, 3, 33, 101
natural, 2, 34, 43, 55, 57, 60, 64, 67, 68, 70, 86
Natural Resources Conservation Service, 35
Navy, 3
Nebraska, 11, 14, 21, 22, 25, 33, 43, 91, 107
Netherlands, 31
network, xv, 80, 84
neurodegenerative disease, 84, 88
neurological disease, 18
neurotoxic, x, 11, 85
neurotoxicity, 62
neurotoxins, 9, 84, 88
Nevada, 43, 108
New England, 88
New Mexico, 8, 25, 33, 97
New Orleans, 24
New York, 11, 20, 43, 44, 63, 68, 83, 87, 89, 92, 108, 109
NHC, xviii, 76, 77, 78, 79, 104
NIH, 4
nitrogen, 11, 16, 34, 35, 91, 93, 101, 102
nitrogen fixation, 34, 101, 102
NOAA, xvii, xviii, 5, 11, 20, 31, 32, 34, 38, 41, 42, 43, 44, 45, 46, 47, 49, 50, 58, 59, 74, 79, 81, 100, 104
non-profit, 74, 77
nontoxic, x, 17, 18, 71
North America, 78, 92
North Carolina, xviii, 4, 25, 26, 74, 98, 108
NRM, 100
nutrient, ix, xiv, 10, 16, 24, 25, 34, 67, 69, 85, 90, 93, 99
nutritional supplements, 39, 40, 66

O

Ohio, 14
Oklahoma, 8, 25, 29, 88
olfactory, 9
Oncology, 87
online, 81
Oregon, 19, 20, 33, 39, 43, 53, 74, 108
organic, 11, 25, 34, 37, 102
organic compounds, 37
organic matter, 11, 34
organism, 9, 10, 59, 60
oxidation, 38, 106
oxygen, x, 7, 18
ozone, 38, 41, 106

P

paralysis, 12
Paris, 84
partnership, 47, 50, 98, 111
pathogens, 25
pathology, 90
pathways, 60
patients, 13, 14
PCR, 32
peer, xv, 4, 50, 101
peptide, 87, 89, 91
perceptions, xiv, 23, 68, 73
pets, 19, 42, 43
Pfiesteria, 45, 108
Phoenix, 38, 105
phosphorus, 11, 16, 35, 93
physiological, 63, 67
phytoplankton, 16, 40, 90, 93, 102
plankton, 39, 71, 84
planning, 31, 48, 53, 75, 80
plants, 36, 99
platforms, 70
poisoning, 19, 91
Poland, 31
pollution, ix, 25, 36
Portugal, 31
Potomac River, 26, 27, 41, 107

precipitation, 67
predators, 19
prediction, 11, 30, 42, 56, 59, 73, 75
predictive model, xiv, 34, 35, 67
prevention, xiv, 4, 11, 30, 34, 36, 45, 54, 56, 59, 68, 69, 70, 75, 92, 108
preventive, 68
prices, 17, 23
private, xi, 19, 43, 50, 75, 108
probe, 45, 100
production, ix, 11, 14, 17, 25, 26, 34, 45, 49, 56, 60, 67, 71, 94, 95, 102
profit, 74, 77
program, xi, xiv, xv, 1, 5, 11, 27, 31, 32, 33, 38, 48, 50, 74, 75, 105, 108
proliferation, 9, 34
proteins, 62
protocols, 44, 57
public health, xii, 17, 21, 22, 27, 32, 43, 45, 46, 49, 78, 83, 92, 98, 99, 108
purification, 38, 111

Q

quality control, 39, 57
quality improvement, 16
quality of life, 23

R

radiation, 41
range, x, 41, 63, 64, 73
recreation, xi, 23, 66, 70
recreational, ix, 2, 7, 10, 13, 14, 19, 21, 22, 23, 24, 27, 33, 35, 43, 46, 49, 61, 66, 68, 73, 75, 89, 92, 95, 98, 107, 108
regulations, x, 15, 36, 72
relationships, 47, 54, 61, 67, 72
remediation, 2
remote sensing, 40
reproduction, 47
reptiles, 98
Research and Development, 34, 83, 84, 100
research funding, 30

reservoirs, 8, 22, 33, 36, 39, 40, 41, 45, 87, 92, 96, 97, 101
residues, 42
resolution, 94
resources, xv, 14, 43, 50, 70, 73, 80, 110
respiratory, 9, 12
revenue, 23, 28
rice field, 41, 93
risk, xii, xiii, xiv, 4, 22, 40, 47, 55, 58, 60, 61, 62, 63, 64, 66, 70, 78, 81, 84, 101, 111, 112
risk assessment, xii, xiii, xiv, 4, 55, 58, 60, 61, 62, 64, 70, 78, 81, 101
river basins, 28, 33
rivers, 10, 27, 28, 33, 36, 92
robotic, 33
runoff, 25

S

safe drinking water, 36
safeguard, 56, 73
safety, 18, 23, 32, 39, 40, 43, 54, 73, 99, 109
saline, 4, 7, 28
salinity, 11, 24, 25
sampling, 21, 22, 32, 33, 57, 58, 65, 94, 102, 106, 109
satellite, 42, 49, 94
scientific community, 104
scientific knowledge, xi, 4
scientific understanding, xiv
SDWA, xviii, 2, 48, 69
seafood, 73, 99
Seattle, 16
sedimentation, 37, 38
sediments, 64
sensors, xii, 54, 57, 72
separation, 44, 101
services, 70
severity, xiii, 65, 78
sewage, 16
sharing, 50, 80
sheep, 19
shellfish, 33, 100

shores, 32
short run, xiii, 65
short-term, 50, 63
signs, 10, 43
sociocultural, xi, 23
socioeconomic, 59
South Carolina, 25, 74, 98
Spain, 10, 31
spatial, 31, 32, 56, 65, 66, 95
species, ix, 4, 8, 9, 11, 25, 28, 29, 32, 34, 36, 47, 63, 64, 65, 67, 69, 72, 97, 98, 102
spectrum, 40
stability, 11, 60
stages, 68
stakeholder, xiv
stakeholders, 19, 73, 75, 81
standards, xiv, 36, 44, 48, 56, 70, 96, 111
starvation, 19
strains, 32, 45, 67, 71, 94, 95, 102
strategies, xii, xiv, 4, 6, 9, 16, 30, 34, 39, 45, 46, 52, 54, 56, 67, 68, 69, 70, 72, 73, 74, 75, 82, 110, 111
streams, 10, 33, 43
stressors, 19, 47, 86
substances, 112
sulfate, 13, 40, 41
sunlight, x, 19
supplements, 40, 55, 99
supply, 14, 20, 45, 88
surface water, 13, 35, 42, 69, 93
surveillance, 45, 72
survival, 25, 47
Sweden, 14, 83
swimmers, 22
Switzerland, 31, 92
symptoms, 17, 98
synthesis, 45, 55, 60

T

task force, 43
taste, 7, 8, 9, 14, 17, 23, 36, 38, 41, 97, 105
taxa, 7, 8, 10, 25, 63, 85, 95, 102
taxonomic, 33, 56, 72
taxonomy, 57, 71, 73

temperature, 11, 37, 41, 67
temporal, 31, 32, 56, 65, 66, 95
Tennessee, 54, 64
Texas, xi, xiii, xviii, xix, 8, 24, 25, 28, 33, 41, 43, 86, 89, 92, 97, 102, 109
threat, x, xii, 15, 17, 19, 45, 59, 79, 104
threatened, 25, 28
threats, 23, 27, 30, 32, 43, 50, 73, 95, 106
time periods, xiii, 63
tissue, 60, 95
titanium, 38, 101
titanium dioxide, 38, 101
tourism, 23, 28
toxic, xii, xiii, 7, 13, 17, 18, 19, 20, 22, 25, 26, 27, 28, 29, 31, 32, 33, 40, 46, 47, 60, 62, 63, 64, 65, 69, 71, 83, 84, 89, 95, 97, 112
toxic substances, 112
toxicity, xiv, 2, 11, 19, 22, 37, 40, 45, 46, 47, 48, 50, 61, 62, 67, 88, 90, 92, 97, 99
toxicological, 70
toxicology, 46, 55, 71, 78, 81, 89
toxicology studies, 55
training, 14, 33, 70, 71, 73, 79, 80, 81, 96, 110
transfer, 55, 64
transformation, 64
transformations, 91
transport, xiv, 41, 50, 67
treatment methods, 38
tribal, 22, 73, 75, 105
tribal lands, 22
tropical areas, 15
trout, 28, 90, 91
tumors, 12
Turkey, 82

U

U.S. Department of Agriculture (USDA), 4
U.S. Geological Survey, xix, 4, 97
ultrasound, 40, 69
ultraviolet, 37, 41
uncertainty, 62, 63
UNESCO, xviii, 80, 110

Index

United Kingdom, 31, 89
United Nations, xviii, 80
United States, ix, x, xi, 2, 4, 6, 7, 9, 10, 11, 14, 15, 16, 17, 23, 25, 29, 30, 31, 34, 44, 46, 54, 59, 74, 75, 78, 80, 81, 82, 86, 95, 96, 103
USDA, xix, 4, 35, 40, 41, 58, 59

V

validation, 81, 110
values, ix, 7, 47
variability, 95
variables, 41, 67
Vermont, 20, 21, 33, 43, 109
viruses, 67
visible, 10, 42, 43

W

wastewater, 26, 35, 36, 93
water quality, ix, 7, 32, 33, 34, 35, 81, 92, 97, 98, 109
water resources, 80
water supplies, 38, 40, 69
waterfowl, 20
watersheds, 34, 36, 98
waterways, 37, 99
weather prediction, 42
web, 11, 19, 30, 43, 45, 46, 48, 61, 79, 94, 98, 105, 106, 107, 108
web-based, 45, 46, 98
websites, 5, 43
wetlands, 33, 35, 93, 97
WHO, xix, 4, 10, 11, 13, 17, 18, 20, 24, 27, 32, 92, 112
wild animals, 91
wildlife, 19, 30, 33, 47, 98
winter, 28
Wisconsin, 42, 78, 83, 109
wood, 36
working groups, 81
World Health Organization, x, xix, 17, 92, 112
Wyoming, 25

Y

yield, 71

Z

zebrafish, 46, 99
zooplankton, 89

DATE DUE
Rtnd - AM AUG -3 2012

TN: **4387294**
Pieces: **1**
ILL: 91958450
NXW   07/30/12